Megaprojects and Risk

Megaprojects and Risk provides the first detailed examination of the phenomenon of megaprojects. It is a fascinating account of how the promoters of multibillion-dollar megaprojects systematically and self-servingly misinform parliaments, the public and the media in order to get projects approved and built. It shows, in unusual depth, how the formula for approval is an unhealthy cocktail of underestimated costs, overestimated revenues, undervalued environmental impacts and overvalued economic development effects. This results in projects that are extremely risky, but where the risk is concealed from MPs, taxpayers and investors. The authors not only explore the problems but also suggest practical solutions drawing on theory and hard, scientific evidence from the several hundred projects in twenty nations that illustrate the book. Accessibly written, it will be essential reading in its field for students, scholars, planners, economists, auditors, politicians, journalists and interested citizens.

BENT FLYVBJERG is Professor in the Department of Development and Planning at Aalborg University, Denmark, and author of the highly successful *Making Social Science Matter* (Cambridge, 2001) and *Rationality and Power* (1998).

NILS BRUZELIUS is Associate Professor at Stockholm University and an independent consultant on transport and planning.

WERNER ROTHENGATTER is Head of the Institute of Economic Policy Research and of the Unit on Transport and Communication at the University of Karlsruhe, Germany.

Megaprojects and Risk

An Anatomy of Ambition

Bent Flyvbjerg
Nils Bruzelius
Werner Rothengatter

CAMBRIDGE
UNIVERSITY PRESS

CAMBRIDGE
UNIVERSITY PRESS

University Printing House, Cambridge CB2 8BS, United Kingdom

Published in the United States of America by Cambridge University Press, New York

Cambridge University Press is part of the University of Cambridge.

It furthers the University's mission by disseminating knowledge in the pursuit of education, learning and research at the highest international levels of excellence.

www.cambridge.org
Information on this title: www.cambridge.org/9780521009461

© Bent Flyvbjerg, Nils Bruzelius, Werner Rothengatter 2003

This publication is in copyright. Subject to statutory exception and to the provisions of relevant collective licensing agreements, no reproduction of any part may take place without the written permission of Cambridge University Press.

First published 2003
13th printing 2013

Printed in the United Kingdom by Clays, St Ives plc.

A catalogue record for this publication is available from the British Library

Library of Congress Cataloguing in Publication data
Flyvbjerg, Bent.
Megaprojects and risk: an anatomy of ambition / by Bent Flyvbjerg, Nils Bruzelius, Werner Rothengatter.
  p. cm.
Includes bibliographical references and index.
ISBN 0 521 80420 5 – ISBN 0 521 00946 4 (pb.)
   1. Project management.   2. Risk management.   I. Bruzelius, Nils.
II. Rothengatter, Werner.   III. Title.
HD69.P75 F59 2002   658.4'04–dc21   2002074193

ISBN 978-0-521-00946-1 Paperback

Cambridge University Press has no responsibility for the persistence or accuracy of URLs for external or third-party internet websites referred to in this publication, and does not guarantee that any content on such websites is, or will remain, accurate or appropriate. Information regarding prices, travel timetables and other factual information given in this work are correct at the time of first printing but Cambridge University Press does not guarantee the accuracy of such information thereafter.

Contents

	List of figures	*page* vi
	List of tables	vii
	Acknowledgements	viii
1	The megaprojects paradox	1
2	A calamitous history of cost overrun	11
3	The demand for megaprojects	22
4	Substance and spin in megaproject economics	32
5	Environmental impacts and risks	49
6	Regional and economic growth effects	65
7	Dealing with risk	73
8	Conventional megaproject development	86
9	Lessons of privatisation	92
10	Four instruments of accountability	107
11	Accountable megaproject decision making	125
12	Beyond the megaprojects paradox	136
	Appendix. Risk and accountability at work: a case study	143
	Notes	152
	Bibliography	180
	Index	201

Figures

2.1	Construction cost overruns for the Channel tunnel, Great Belt link and Øresund link	*page* 12
2.2	Year-by-year development in forecasts of construction costs and road traffic for the Great Belt and Øresund links	13
2.3	Inaccuracies of cost estimates for rail, fixed links and roads in 258 projects	17
2.4	A century of cost overrun in 111 projects	18
3.1	Forecast and actual traffic for the Channel tunnel	23
3.2	Inaccuracies of traffic forecasts in 210 transport infrastructure projects	27
4.1	Eurotunnel share prices, 1987–2001	33
5.1	Impact of the Great Belt link on common mussels before, during and after construction	59
7.1	The risk management process	82
9.1	The typical BOT (build-operate-transfer) approach to megaproject development	94
10.1	A stakeholder-based approach to decision making	113
10.2	French approach for ensuring transparency in megaproject decision making	114
10.3	Performance specifications and the stakeholder approach in combination	116

Tables

		page
II.i	Examples of construction cost overruns in large transport projects	14
II.ii	Spectacular projects with spectacular cost overruns	19
II.iii	Construction cost development in four privately owned transport infrastructure projects	20
III.i	Examples of projects that have experienced serious revenue/benefit problems	25
IV.i	Cost overrun and ridership for twelve urban rail projects	38
IV.ii	Examples of projects that have experienced viability problems	42
VIII.i	Steps in the conventional approach to project development	87
IX.i	Cost and traffic development in four privately owned transport infrastructure projects	101
XI.i	Alternative 1: the concession approach to project development	126
XI.ii	Alternative 2: the state-owned enterprise approach to project development	127

Acknowledgements

We wish to thank the people and organisations who helped make this book possible. Special thanks must be given to Patrick Ponsolle and John Noulton, of Eurotunnel, Mogens Bundgaard-Nielsen, of Sund & Bælt Holding, and Ole Zacchi, of the Danish Ministry of Transport. Not only did they and their staff help us with data for the book's case studies, they also gave critical comments on an earlier version of the book's manuscript.

We also wish to thank Martin Wachs, of the University of California at Berkeley, and Don Pickrell, of the Volpe National Transportation Systems Center at Cambridge, Massachusetts, for their comments on our analysis of cost overrun. Per Homann Jespersen, of Roskilde University, provided helpful input to our considerations regarding environmental impacts and risks. Roger Vickerman, of the University of Kent at Canterbury, gave valuable comments on the chapter about regional and economic growth effects. Thanks are due as well to the following colleagues for their kind help at various stages in the research and writing process: Jim Bohman, Irene Christiansen, John Dryzek, Raphael Fischler, Ralph Gakenheimer, Maarten Hajer, Mette Skamris Holm, Andy Jamison, Bill Keith, Finn Kjærsdam, Mary Rose Liverani, Kim Lynge Nielsen, Tim Richardson, Yvonne Rydin, Ed Soja, Michael Storper, Andy Thornley, Jim Throgmorton and Alan Wolfe. Two anonymous Cambridge University Press reviewers provided highly useful comments for preparing the final version of the typescript.

The transport sector and its institutions are hardly in the vanguard regarding freedom of information. In some cases we were unable, using the formal channels for information gathering, to get the data and in-depth information we needed to write the book the way we wanted to write it. We are grateful to those bold individuals who, when the formal channels dried up, found informal ways of furnishing us with the information we lacked. We mention no names for obvious reasons.

Lilli Glad expertly transformed our drafts into readable manuscripts. Anni Busk Nielsen provided precious help in acquiring the literature on which the study is based. The research and the book were made possible by generous grants from the Danish Transport Council and Aalborg University. Finally, we wish to thank our editor at Cambridge University Press, Sarah Caro, who provided valuable help in seeing the book through the printing process. Bent Flyvbjerg was teamleader for the research on which the book is based and is principal author of the book. We apologise to anyone we have forgotten to mention here. Responsibility for errors or omissions in this book remains ours alone.

1 The megaprojects paradox

A new animal

Wherever we go in the world, we are confronted with a new political and physical animal: the multibillion-dollar mega infrastructure project. In Europe we have the Channel tunnel, the Øresund bridge between Denmark and Sweden, the Vasco da Gama bridge in Portugal, the German MAGLEV train between Berlin and Hamburg, the creation of an interconnected high-speed rail network for all of Europe, cross-national motorway systems, the Alp tunnels, the fixed link across the Baltic Sea between Germany and Denmark, plans for airports to become gateways to Europe, enormous investments in new freight container harbours, DM 200 billion worth of transport infrastructure projects related to German unification alone, links across the straits of Gibraltar and Messina, the world's longest road tunnel in Norway, not to speak of new and extended telecommunications networks, systems of cross-border pipelines for transport of oil and gas, and cross-national electrical power networks to meet the growing demand in an emerging European energy market. It seems as if every country, and pair of neighbouring countries, is in the business of promoting this new animal, the megaproject, on the European policy-making scene. And the European Union, with its grand scheme for creating so-called 'Trans-European Networks', is an ardent supporter and even initiator of such projects, just as it is the driving force in creating the regulatory, and de-regulatory, regimes that are meant to make the projects viable.[1]

The situation is similar in industrialised and industrialising countries in other parts of the world, from Asia to the Americas. There is, for example, Hong Kong's Chek Lap Kok airport, China's Quinling tunnel, the Akashi Kaikyo bridge in Japan, Sydney's harbour tunnel, Malaysia's North–South Expressway, Thailand's Second Stage Expressway, and proposals for an integrated Eurasian transport network. In the Americas there is Boston's 'Big Dig', freeways and railways in California, Denver's new international airport, Canada's Confederation bridge, the São Paulo–Buenos Aires Superhighway, the Bi-Oceanic highway right across South America from the Atlantic to the Pacific, and the Venezuela–Brazil highway. Even a proposed US$50 billion project to link the USA and Russia across the Bering Strait – the 'biggest project in history', according to its promoters – is not missing in the megaproject scheme of things.[2] Outside the field of transport infrastructure there is the Three Gorges dam in China, Russia's natural gas pipelines, the Pergau dam in Malaysia, flood control in Bangladesh, the Bolivia–Brazil gas pipeline, the Venezuela–Brazil power line and, again and everywhere, the ultimate megaproject, the Internet with associated infrastructure and telecommunications projects.

> Megaprojects form part of a remarkably coherent story, the 'Great War of Independence from Space'.

Zero-friction society

Megaprojects form part of a remarkably coherent story. Sociologist Zygmunt Bauman perceptively calls it the 'Great War of Independence from Space', and he sees the resulting new mobility as the most powerful and most coveted stratifying factor in contemporary society.[3] Paul Virilio speaks of the 'end of geography' while others talk of the 'death of distance'.[4] Bill Gates, founder and chair of Microsoft Corporation, has dubbed the phenomenon 'frictionless capitalism' and sees it as a novel stage in capitalist evolution.[5] When Microsoft and Gates single out a concept or a product, one is well advised to pay attention. 'Frictionless society' may sound like an advertiser's slogan in the context of its usage. It is not. The term signifies a qualitatively different stage of social and economic development.

In this development 'infrastructure' has become a catchword on a par with 'technology'. Infrastructure has rapidly moved from being a simple precondition for production and consumption to being at the very core of these activities, with just-in-time delivery and instant Internet access being two spectacular examples of this. Infrastructure is the great space shrinker, and power, wealth and status increasingly belong to those who know how to shrink space, or know how to benefit from space being shrunk.[6]

Today infrastructure plays a key role in nothing less than the creation of what many see as a new world order where people, goods, energy, information and money move about with unprecedented ease. Here the politics of distance is the elimination of distance. The name of utopia is Zero-Friction Society. And even if we can never achieve utopian frictionlessness, we may get close, as is currently happening with the spread of the Internet. Modern humans clearly have a preference for independence from space and are consistently undercutting the friction of distance by building more and improved infrastructure for transport, including telecommunications and energy.

Megaprojects are central to the new politics of distance because infrastructure is increasingly being built as megaprojects. Thus the past decade has seen a sharp increase in the magnitude and frequency of major infrastructure projects, supported by a mixture of national and supranational government, private capital and development banks.

> Many projects have strikingly poor performance records in terms of economy, environment and public support.

Performance paradox

There is a paradox here, however. At the same time as many more and much larger infrastructure projects are being proposed and built around the world, it is becoming clear that many such projects have strikingly poor performance records in terms of economy, environment and public support.[7] Cost overruns and lower-than-predicted revenues frequently place project viability at risk and redefine projects that were initially promoted as effective vehicles to economic growth as possible obstacles to such growth. The Channel tunnel, opened in 1994 at a construction cost of £4.7 billion, is a case in point, with several near-bankruptcies caused by construction cost overruns of 80 per cent, financing costs that are 140 per cent higher than those forecast and revenues less than half of those projected (see Chapters 2–4). The cost overrun for Denver's US$5 billion new international airport, opened in 1995, was close to 200 per cent and passenger traffic in the opening year was only half of that projected. Operating problems with Hong Kong's new US$20 billion Chek Lap Kok airport, which opened in 1998, initially caused havoc not only to costs and revenues at the airport; the problems spread to the Hong Kong economy as such with negative effects on growth in gross domestic product.[8] After nine months of operations, *The Economist* dubbed the airport a 'fiasco', said to have cost the Hong Kong economy US$600 million.[9] The fiasco may have been only a start-up problem, albeit an expensive one, but it is the type of expense that is rarely taken into account when planning megaprojects.

Some may argue that in the long term cost overruns do not really matter and that most monumental projects that excite the world's imagination had large overruns. This line of argument is too facile, however. The physical and economic scale of today's megaprojects is such that whole nations may be affected in both the medium and long term by the success or failure of just a single project. As observed by Edward Merrow in a RAND study of megaprojects:

> Such enormous sums of money ride on the success of megaprojects that company balance sheets and even government balance-of-payments accounts can be affected for years by the outcomes... The success of these projects is so important to their sponsors that firms and even governments can collapse when they fail.[10]

Even for a large country such as China, analysts warn that the economic ramifications of an individual megaproject such as the Three Gorges dam 'could likely hinder the economic viability of the country as a whole'.[11] Stated in more general terms, the Oxford-based Major Projects Association, an organisation of contractors, consultants, banks and others interested in megaproject development, in a recent publication speaks of the 'calamitous history of previous cost overruns of very large projects in the public sector'. In another study sponsored by the Association the conclusion is, 'too many projects proceed that should not have done'.[12] We would add to this that regarding cost overruns there is no indication that the calamity identified by the Major Projects Association is limited to the public sector. Private sector cost overruns are also common.

For environmental and social effects of projects, one similarly finds that such effects often have not been taken into account during project development, or they have been severely miscalculated.[13] In Scandinavia, promoters of the Øresund and Great Belt links at first tried to ignore or downplay environmental issues, but were eventually forced by environmental groups and public protest to accept such issues on the decision-making agenda (see Chapter 5). In Germany, high-speed rail projects have been criticised for not considering environmental disruption. Dams are routinely criticised for the same thing. However, environmental problems that are not taken into account during project preparation tend to surface during construction and operations; and such problems often destabilise habitats, communities and megaprojects themselves, if not dealt with carefully. Moreover, positive regional development effects, typically much touted by project promoters to gain political acceptance for their projects, repeatedly turn out to be non-measurable, insignificant or even negative (see Chapter 6).

In consequence, the cost–benefit analyses, financial analyses and environmental and social impact statements that are routinely carried out as part of megaproject preparation are called into question, criticised and denounced more often and more dramatically than analyses in any other professional field we know. Megaproject development today is not a field of what has been called 'honest numbers'.[14] It is a field where you will see one group of professionals calling the work of another not only 'biased' and 'seriously flawed' but a 'grave embarrassment' to the profession.[15] And that is when things have not yet turned unfriendly. In more antagonistic situations the words used in the mud-slinging accompanying many megaprojects are 'deception', 'manipulation' and even 'lies' and 'prostitution'.[16] Whether we like it or not, megaproject development is currently a field where little can be trusted, not even – some would say especially not – numbers produced by analysts.

Finally, project promoters often avoid and violate established practices of good governance, transparency and participation in political and administrative decision making, either out of ignorance or because they see such practices as counterproductive to getting projects started. Civil society does not have the same say in this arena of public life as it does in others; citizens are typically kept at a substantial distance from megaproject decision making. In some countries this state of affairs may be slowly changing, but so far megaprojects often come draped in a politics of mistrust. People fear that the political inequality in access to decision-making processes will lead to an unequal distribution of risks, burdens and benefits from projects.[17] The general public is often sceptical or negative towards projects; citizens and interest groups orchestrate hostile protests; and occasionally secret underground groups even encourage or carry out downright sabotage on projects, though this is not much talked about in public for fear of inciting others to similar guerrilla activities.[18] Scandinavians, who like other people around the world have experienced the construction of one megaproject after another during the past decade, have coined a term to describe the lack in megaproject decision making of accustomed transparency and involvement of civil society: 'democracy deficit'. The fact that a special term has come into popular usage to describe what is going on in megaproject decision making is indicative of the extent to which large groups in the population see the current state of affairs as unsatisfactory.

Civil society does not have the same say in this arena of public life as it does in others. Megaprojects often come draped in a politics of mistrust.

Risk, democracy and power

The megaprojects paradox consists in the irony that more and more megaprojects are built despite the poor performance record of many projects. In this book we link the idea of megaprojects with the idea of risk and we identify the main causes of the megaprojects paradox to be inadequate deliberation about risk and lack of accountability in the project decision-making process. We then proceed to propose ways out of the paradox. We will show that in terms of risk, most appraisals of megaprojects assume, or pretend to assume, that infrastructure policies and projects exist in a predictable Newtonian world of cause and effect where things go according to plan. In reality, the world of megaproject preparation and implementation is a highly risky one where things happen only with a certain probability and rarely turn out as originally intended.

Sociologists such as Ulrich Beck and Anthony Giddens have argued that in modern society risk has increasingly become central to all aspects of human affairs; that we live in a 'risk society' where deliberation about social, economic, political and environmental issues is bound to fail if it does not take risk into account.[19] If this diagnosis is correct – and we will argue that for megaprojects it is – then it is untenable to continue to act as if risk does not exist or to underestimate risk in a field as costly and consequential as megaproject development.

The Beck–Giddens approach to risk society is our point of departure for understanding risk and its particular relevance to modern society. Yet this approach does not take us far enough in the direction we want to go. The problem with Beck, Giddens and related theories is that they use risk mainly as a metaphor for mature modernity. We want to proceed beyond the level of symbol and theory to use risk as an analytic frame and guide for actual decision making. We will do this by developing a set of ideas of how risk assessment and risk management may be used as vehicles for governing risk.[20] In the words of Silvio Funtowicz and Jerome Ravetz, where facts are uncertain, decision-stakes high and values in dispute, risk assessment must be at the heart of decision making.[21] A growing number of society's decision areas meet these criteria. Megaproject development is one of them.

We do not believe risk can be eliminated from risk society. We believe, however, that risk may be acknowledged much more explicitly and managed a great deal better, with more accountability, than is typically the case today. Like Ortwin Renn, Thomas Webler and others, we hold that risk assessment and management should involve citizens and stakeholders to reflect their experience and expertise, in addition to including the usual suspects, namely government experts, administrators and politicians.[22]

We here define stakeholders as key institutional actors, such as NGOs, various levels of government, industrial interests, scientific and technical expertise and the media. Some of these stakeholder groups will claim to be speaking legitimately on behalf of the public good and some, but not all of them, will be doing so. Given that such stakeholders do not always adequately represent publics, we recognise the need, on both democratic and pragmatic grounds, to properly involve publics in decision making. Such involvement should take place in carefully designed deliberative processes from the beginning and throughout large-scale projects.[23] Like Renn and Webler, we believe that one should go as far as possible with the participatory and deliberative approach in including publics and stakeholders and that the result will be decisions about risk that are better informed and more democratic.

We find, nevertheless, that deliberative approaches to risk, based as they are on communicative rationality and the goodwill of participants, can take us only some of the way towards better decisions and will frequently fail for megaprojects.[24] This is so because the interests and power relations involved in megaprojects are typically very strong, which is easy to understand given the enormous sums of money at stake, the many jobs, the environmental impacts, the national prestige, and so on. Communicative and deliberative approaches work well as ideals and evaluative yardsticks for decision making, but they are quite defenceless in the face of power.[25] And power play, instead of commitment to deliberative ideals, is often what characterises megaproject development. In addition to deliberative processes, we also focus, therefore, on how power relations and outcomes may be influenced and balanced by reforming the institutional arrangements that form the context of megaproject decision making.[26]

Based on this approach to risk, it is an essential notion of the book that good decision making is a question not only of better and more rational information and communication, but also of institutional arrangements that promote accountability, and especially accountability towards risk. We see accountability as being a question not just about periodic elections, but also about a continuing dialogue between civil society and policy makers and about institutions holding each other accountable through appropriate checks and balances.[27] Thus we replace the conventional decisionistic approach to megaproject development with a more current institutionalistic one centred on the practices and rules that comprise risk and accountability.[28] We also hold that our approach must be based on actual experience from concrete projects. The purpose is to ensure a realistic understanding of the issues at hand as well as proposals that are practically desirable and possible to implement.

A brief overview

We build our case for a new approach to megaproject decision making in two main steps. In the first half of the book, we identify the weaknesses of the conventional approach to megaproject development. By so doing we argue that a different approach is needed. Our critique of the conventional approach is proactive; from the critique we tease out the problems, namely problems that need to be embraced by an alternative approach. In the second half of the book, we argue empirically and theoretically how the weaknesses of the conventional approach can be overcome by emphasising risk, institutional issues and accountability. Finally, an example for readers with a practical bent is included in the appendix, which shows how our approach to megaproject decision making was employed for a specific project with which we have been involved as advisers to the Danish government, namely the proposed Baltic Sea link connecting Germany and Denmark across Fehmarn Belt, one of the largest cross-national infrastructure projects in the world.

Throughout the book we illustrate major points on the basis of in-depth case studies of three recent megaprojects that form part of the so-called Trans-European Transport Network sponsored by the European Union and national governments:

(1) the Channel tunnel, also known as the 'Chunnel', between France and the UK, which opened in 1994 and is the longest underwater rail tunnel in Europe;
(2) the Great Belt link, opened in 1997–98, connecting East Denmark with continental Europe, and including the longest suspension bridge in Europe plus the second longest underwater rail tunnel; and finally
(3) the Øresund link between Sweden and Denmark, which opened in 2000, and which connects the rest of Scandinavia with continental Europe.

The three case studies are supplemented by material from a large number of other major projects, mainly from the field of transport infrastructure, but also from other fields such as information technology, power plants, water projects, oil and gas extraction projects and aerospace projects. The economics and politics of building a bridge or an airport are surely different on many points from those of space exploration, water management, or providing global access to the Internet. But despite such differences, our data show that there are also important similarities, for instance regarding cost overrun and financial risk, where we find a remarkably similar pattern across different project types. We argue that the measures of accountability that are necessary for detecting and curbing systematic cost underestimation, benefit overestimation and other risks are quite similar across projects. Thus, even though the main focus of the book is the development of mega transport infrastructure projects, the approach developed is relevant for other types of megaproject as well.

Our case studies and other data cover both public and private sector projects. We argue that for megaprojects there is no simple formula for the government–business divide. Megaprojects are so complicated that by nature they are essentially hybrid. This is the case even for projects that are considered fully private, for instance the Channel tunnel, because the sheer complexity and potential impacts of a megaproject dictate deep public-sector involvement for many issues, for instance regarding safety and environment. Thus public–private collaboration is crucial, even for private-sector projects. The question is not whether such collaboration should take place but how. In Chapters 9 and 10 we address this question and redraw the borderlines of public and private involvement in megaproject development with a view to improving governance of risk.

By linking the idea of megaprojects with the idea of risk we hope to broaden the scope of the risk literature and to attract attention to this topic. As far as we are aware, no other study does this today. In writing the book, we have aimed at an interdisciplinary audience of students and scholars in the social and decision-making sciences with an interest in risk, public policy and planning, ranging from sociology and social policy to political science and public policy to public administration, management and planning. Policy makers, administrators and planners are also an important target group for the book, as are consultants, auditors and other practitioners working with megaproject development. We maintain that governments and developers who continue to ignore the type of knowledge and proposals presented here do so at their own peril. Megaprojects are increasingly becoming highly public and intensely politicised ventures drawing substantial international attention with much potential for generating negative publicity.

The Three Gorges dam mentioned above is a case in point. So is the 650 km Myanmar–Thailand natural gas pipeline and maintenance road, built through pristine natural forests and habitats. Lonely Planet, the world's leading travel guidebook publisher, decided to print, up front in its best-selling Thailand guide, a highly visible protest against the pipeline which called the actions of both the Thai government and named transnational companies, such as American Unocal and French Total, a 'scam', 'shameful' and a *'fait accompli'*.[29] Lonely Planet encourages the reader to join the protests against the project and lists – thorough as always – three addresses and telephone and fax numbers where that is possible. This is hardly how the Tourist Authority of Thailand would have preferred to present the country to its visitors, nor is it the type of publicity that transnational corporations opt for, if they have a choice. Our point is they do: there is another way to deal with megaprojects and this book explains what this is.

Finally, though we did not write the book with lay readers primarily in mind, we hope that individuals, communities, activists, media and the general public interested in and affected by megaproject development will find useful insights in the book, for instance regarding the deceptions and power games they are likely to encounter if they get involved with megaprojects. Understanding the anatomy of megaprojects is necessary to be an effective player in project development. And, as mentioned, we see stronger involvement by civil society and stakeholder groups in megaproject decision making as a prerequisite for decisions that are better informed and more democratic.

Theorists of risk society and democracy have recently begun to contemplate the type of practical policy and planning needed for dealing with risk in real-life public deliberation and decision making. 'In risk society', one study concludes, 'public policy requires long-term planning for uncertainty, within a clear framework of principles and evidence to support devolved and flexible decision making. This, in turn, requires the involvement of informed and active citizens, enjoying a mature, adult-to-adult relationship with experts and with politicians. A high-trust democracy: the only way to face a risky future.'[30] In order for this approach to work, the trust in 'high-trust democracy' must be based on, not feet-in-the-air idealism about the merits of democracy, but hard-nosed considerations about risk and democratic accountability. Life will never be risk free, we are happy to report. But risk can be faced in ways much more intelligent than those currently seen. We offer this book as an attempt at fleshing out in practice the type of decision making and democracy called for by theorists of risk and democracy for a specific domain of increasing social, economic and political importance, namely that of megaproject development.

2 A calamitous history of cost overrun

In this and the following chapters we review the experience from a large number of megaprojects, including the Channel tunnel and the links across Great Belt and Øresund, all three multibillion-dollar projects. Although we subject these projects to critical scrutiny, our objective is not to criticise them, even where they have underperformed, but to learn constructively from experience by identifying lessons that may prove useful in improving future decisions regarding megaprojects. Given the large amounts of money spent on major transport infrastructure projects, it is remarkable how little data and research are available that would help answer the two basic questions: (i) whether such projects have the intended effects; and (ii) how the actual viability of such projects compares to projected viability. Therefore, in addition to transport projects, we have found it pertinent to review data and research from other types of infrastructure project and to compare experience from these projects with experience from the transport sector. In this manner, we will review data from several hundred large projects. In this chapter we focus on the costs of megaprojects. In Chapters 3 and 4 we consider the demand for, and viability of, such projects.

> A first step in reducing cost overrun is to acknowledge that a substantial risk for overrun exists and cannot be completely eliminated; but it can be moderated.

The problem of cost overrun

Cost overruns in major transport infrastructure projects are widespread. The difference between actual and estimated investment cost is often 50–100 per cent, and for many projects cost overruns end up threatening project viability. A first step in reducing cost overrun is to acknowledge that a substantial risk of overrun exists and cannot be completely

12   Megaprojects and Risk

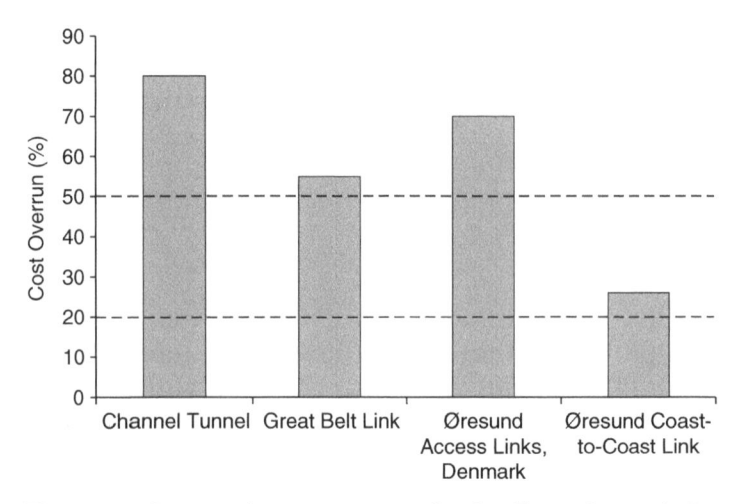

Figure 2.1 Construction cost overruns for the Channel tunnel, Great Belt link and Øresund link (constant prices)

eliminated; but it can be moderated. A next step is to allocate the risk of overrun to those best able to manage it. Both steps will be considered in what follows.

A main cause of overruns is a lack of realism in initial cost estimates. The length and cost of delays are underestimated, contingencies are set too low, changes in project specifications and designs are not sufficiently taken into account, changes in exchange rates between currencies are underestimated or ignored, so is geological risk, and quantity and price changes are undervalued as are expropriation costs and safety and environmental demands. Many major projects also contain a large element of technological innovation with high risk. Such risk tends to translate into cost increases, which often are not adequately accounted for in initial cost estimates. Let us look at cost development in a number of actual projects.

Channel tunnel, Great Belt and Øresund

The Channel tunnel, also known as the 'Chunnel', is the longest underwater rail tunnel in Europe. It opened in 1994 and connects France and the UK. When the Channel Tunnel Treaty was ratified by the French and British parliaments in 1987, total investment costs for this privately financed project were estimated at £2,600 million (1985 prices). Upon completing the project in 1994 actual costs turned out to be £4,650 million (1985 prices) resulting in a cost overrun of 80 per cent. Actual financing costs turned out to be a steep 140 per cent higher than forecast.[1]

A calamitous history of cost overrun 13

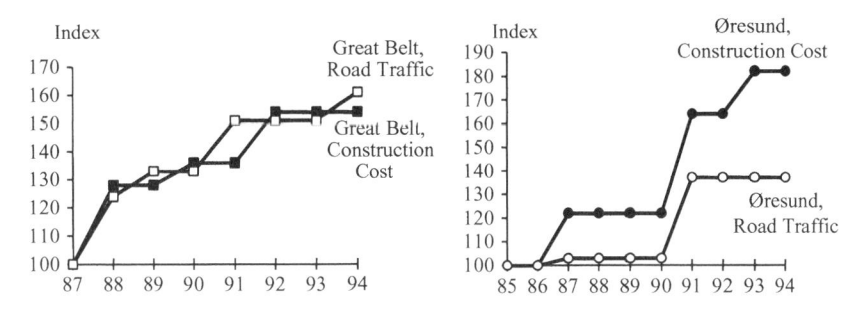

Figure 2.2 Year-by-year development in forecasts of construction costs (constant prices) and road traffic for the Great Belt and Øresund links. The Great Belt link opened in 1997–98; the Øresund link in 2000
Source: Mette K. Skamris and Bent Flyvbjerg, "Inaccuracy of Traffic Forecasts and Cost Estimates on Large Transport Projects", *Transport Policy*, vol. 4, no. 3, 1997, pp. 141–6.

The Great Belt link opened to rail traffic in 1997 and road traffic in 1998. It connects East Denmark with continental Europe and comprises the longest suspension bridge in Europe plus the second longest underwater rail tunnel. When the construction law for the Great Belt link was ratified by the Danish Parliament in 1987, total investment costs were estimated at DKK 13.9 billion (1988 prices). When all construction was finished in 1999, costs had increased by 54 per cent in real terms to DKK 21.4 billion.[2]

The Øresund link between Sweden and Denmark opened in 2000 and is one of the largest cross-national infrastructure projects in the world. It connects Sweden and Norway with continental Europe. When the construction law for the Øresund link was ratified by the Danish Parliament in 1991, total investment costs were estimated at DKK 11.7 billion (1990 prices) for the coast-to-coast project and DKK 3.2 billion for access links on the Danish side.[3] When the Danish access links were completed in 1998, actual costs for these were DKK 5.4 billion (1990 prices), which translates into a cost overrun of 68 per cent.[4] When the coast-to-coast link opened two years later, in 2000, costs for this facility had increased by 26 per cent to DKK 14.8 billion (1990 prices).[5] The Swedish access links have not yet been finished, but the completed parts have an overall cost overrun of 16 per cent. The largest extra cost for the Swedish access links will be an SEK 7 billion rail tunnel in Malmö, the so-called City tunnel. The tunnel was not included in the original scheme for the Øresund link, but it is said to be necessary to link Malmö effectively with the Øresund rail connections to Copenhagen and Denmark. Delays

Table II.i *Examples of construction cost overruns in large transport projects. Constant prices. For patronage figures, see Chapters 3 and 4*

Project	Cost overrun (%)
Boston's artery/tunnel project	196
Humber bridge, UK	175
Boston–Washington–New York rail, USA	130
Great Belt rail tunnel, Denmark	110
A6 Motorway Chapel-en-le-Frith/Whaley bypass, UK	100
Shinkansen Joetsu rail line, Japan	100
Washington metro, USA	85
Channel tunnel, UK, France	80
Karlsruhe–Bretten light rail, Germany	80
Øresund access links, Denmark	70
Mexico City metro line	60
Paris–Auber–Nanterre rail line	60
Tyne and Wear metro, UK	55
Great Belt link, Denmark	54
Øresund coast-to-coast link	26

and cost overruns of 36 per cent have happened even before construction has started. Another extra cost on the Swedish side is the new ten-mile ring motorway around Malmö, which is effectively serving as an access link to the Øresund bridge and which was built about ten years earlier than it would have been without the fixed link to Denmark. Cost development for the Great Belt and Øresund projects are shown in Figure 2.2.[6]

Other transport infrastructure projects

It is often argued that no major projects are alike, and, therefore, that such projects cannot be compared. Indeed, the Channel, Great Belt and Øresund projects are different in many ways. Nevertheless, with respect to cost development, there is a striking similarity between these and other major projects: there is a tendency towards a significant underestimation of costs during project appraisal. This is also the conclusion we draw when we review data from a large number of major transport infrastructure projects, and from other types of project as well. In Table II.i cost overruns for the Channel tunnel, Great Belt link and Øresund links are compared to cost overruns for a number of other transport projects. As we show in Chapters 3 and 4, the problem with cost overrun is exacerbated by the fact that often this problem comes hand in hand with lower-than-estimated revenues. The consequence is projects that are financially risky to the second degree.

Only a few studies exist that rigorously compare forecast with actual costs for more than a few transport infrastructure projects. We have identified four such studies. The first study was carried out by the Auditor-General of Sweden and covered fifteen road and rail projects with a total value of SEK 13 billion (1994 prices).[7] The average capital cost overrun for the eight road projects in the study was 86 per cent, ranging from 2 per cent to 182 per cent, while the average overrun for the seven rail projects was 17 per cent, ranging from −14 to +74 per cent. It should be noted about the Swedish study, however, that two-thirds of the projects in the study were still under construction when the study was carried out. Final costs for these projects may therefore turn out to be higher than the expected final costs quoted by the Swedish Auditor-General.

The second study was carried out by the US Department of Transportation and covered ten US rail transit projects with a total value of US$15.5 billion (1988 prices).[8] The total capital cost overrun of these projects was 61 per cent, ranging from −10 to +106 per cent for the individual projects.

The third study was carried out by the Transport and Road Research Laboratory in the UK and covered twenty-one metro systems in developing and newly industrialised nations each with a value of US$22–165 million (1987 prices).[9] For thirteen of the metros, capital cost overruns could be estimated. Six metros had overruns above 50 per cent, two of these in the range from 100 to 500 per cent. Three metros had overruns in the 20 to 50 per cent range, and the remaining four in the −10 to +20 per cent range.

The fourth and final study was carried out at Aalborg University, Denmark. This is the most comprehensive of the 4 studies and with a sample of 258 projects worth approximately US$90 billion (1995 prices) it is the first study to allow statistically significant conclusions regarding cost overrun in transport infrastructure projects.[10] The project types are bridges, tunnels, highways, freeways, high-speed rail, urban rail and conventional (inter-urban) rail. The projects are located in 20 countries on 5 continents, including both developed and developing nations. The projects were completed between 1927 and 1998. Older projects were included in the sample in order to test whether the accuracy of estimated costs improves over time, that is whether learning regarding cost estimation takes place. Main findings from the Aalborg study are (all highly significant and most likely conservative):

- In nine out of ten transport infrastructure projects costs are underestimated, resulting in cost overrun;
- For rail, actual costs are on the average 45 per cent higher than estimated costs (standard deviation, sd = 38);

- For fixed links (tunnels and bridges), actual costs are on the average 34 per cent higher than estimated costs (sd = 62);
- For roads, actual costs are on the average 20 per cent higher than estimated costs (sd = 30);
- For all project types, actual costs are, on the average, 28 per cent higher than estimated costs (sd = 39);
- Cost underestimation and overrun exist across 20 nations and 5 continents; it appears to be a global phenomenon;
- Cost underestimation and overrun appear to be more pronounced in developing nations than in North America and Europe (data for rail only);
- Cost underestimation and overrun have not decreased over the past seventy years. No learning seems to take place;
- Cost underestimation and overrun cannot be explained by error and seem to be best explained by strategic misrepresentation, namely lying, with a view to getting projects started.

Figure 2.3 shows the inaccuracy of cost estimates in the 258 projects divided into rail, fixed links and roads. This and other studies of cost development in major transport infrastructure projects reveal the same overall pattern: cost overruns above 40 per cent in fixed prices are common, especially for rail projects, and overruns above 80 per cent are not uncommon. Therefore, cost developments for the Chunnel and the Great Belt and Øresund links are not the aberrations they may appear to be at first sight: they are commonplace.

Figure 2.4 shows a plot of cost overrun against year of decision to build for the 111 projects in the sample for which these data are available. The diagram does not seem to indicate an effect from time on cost overrun. Statistical tests corroborate this impression. A test using year of completion instead of year of decision, with data for 246 projects, gives a similar result. We therefore conclude that cost overrun has not decreased over time. Cost overrun today is in the same order of magnitude as it was ten, thirty or seventy years ago. If techniques and skills for estimating costs and avoiding cost overrun in transport infrastructure projects have improved over time, this does not show in the data. No learning seems to take place in this important and highly costly sector of public and private decision making. This seems strange and invites speculation that the persistent existence over time and space and project type of significant and widespread cost overrun is a sign that an equilibrium has been reached: strong incentives and weak disincentives for cost underestimation and thus for cost overrun may have taught project promoters what there is to learn, namely that cost underestimation and overrun pay off. If this is the case, cost overrun must be expected and it must be expected to be intentional.

A calamitous history of cost overrun 17

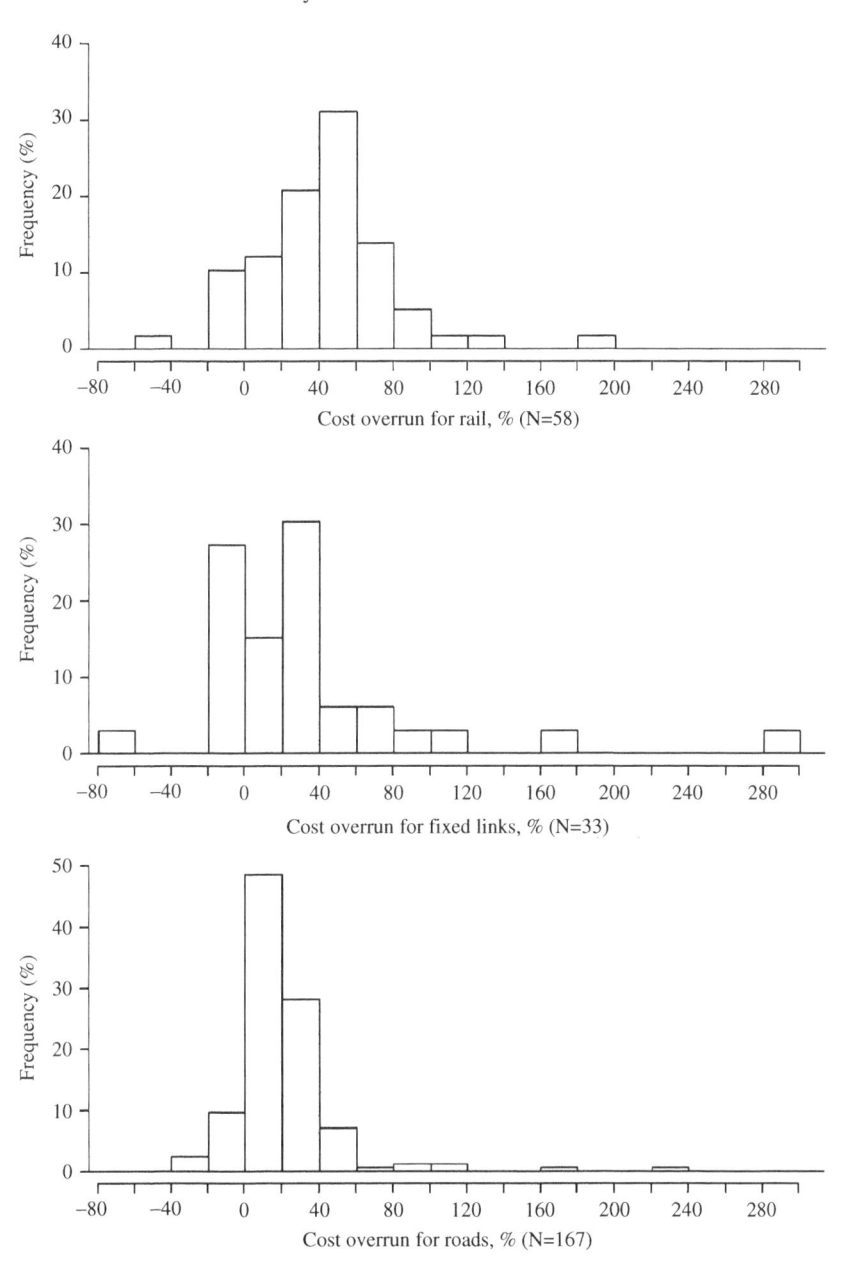

Figure 2.3 Inaccuracies of cost estimates for rail, fixed links and roads in 258 projects. The figure shows the percentage distribution of projects with respect to cost overrun (constant prices)

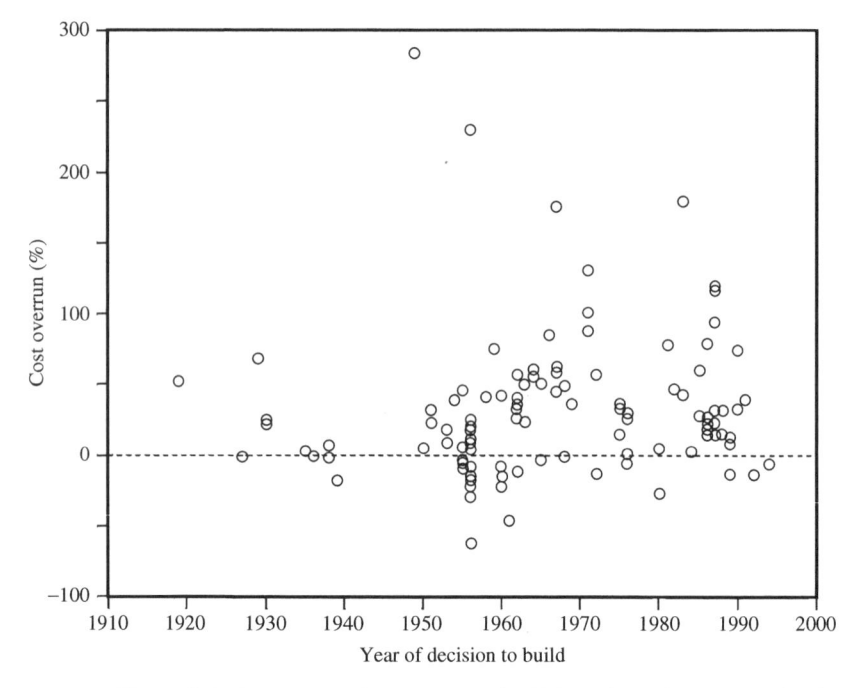

Figure 2.4 A century of cost overrun in 111 projects (constant prices)

Other major projects

In addition to cost data for transport infrastructure projects, we have examined cost data for several hundred other projects, including power plants, dams, water projects, oil and gas extraction projects, information technology systems, aerospace projects and weapons systems.[11] The data show that other types of major project are at least as, if not more, prone to cost overruns as are major transport infrastructure projects.

Among the more spectacular examples of cost overruns are the Sydney Opera House with actual costs approximately fifteen times higher than those projected and the Concorde supersonic aeroplane with twelve times higher costs.[12] The data also indicate that cost overruns for other major projects have neither increased nor decreased historically and that overruns are common in both first and third world countries. When the Suez Canal was completed in 1869 actual construction costs were twenty times higher than the earliest estimated costs and three times higher than the cost estimate for the year before construction began. The Panama Canal, which was completed in 1914, had cost overruns in the range from 70 per cent to 200 per cent.[13]

Table II.ii *Spectacular projects with spectacular cost overruns*

Project	Cost overrun (%)
Suez Canal	1,900
Sydney Opera House	1,400
Concorde supersonic aeroplane	1,100
Panama Canal	200
Brooklyn Bridge	100

Sources: Peter Hall, 'Great Planning Disasters Revisited', p. 3; Robert Summers, 'Cost Estimates as Predictors of Actual Costs: A Statistical Study of Military Developments', in Thomas Marschak *et al.*, eds., *Strategy for R&D: Studies in the Microeconomics of Development* (Berlin: Springer-Verlag, 1967), p. 148; and Mette K. Skamris, 'Economic Appraisal of Large-Scale Transport Infrastructure Investments', Ph.D dissertation (Aalborg: Aalborg University, 2000).

In sum, the phenomenon of cost overrun appears to be characteristic not only of transportation projects but of projects in other fields as well (see Table II.ii).

The *'what'* and *'that'* of cost overrun

As mentioned in Chapter 1, the Major Projects Association in a recent publication speaks of 'the calamitous history of previous cost overruns of very large projects in the public sector'.[14] With the privately owned, privately financed and privately operated Channel tunnel in mind, one must add that there is no clear indication that cost overruns for major projects are a singular public-sector phenomenon or that overruns will be eliminated by simply placing projects in the private sector, even if this may help induce discipline and accountability in major projects as we shall see below (see Table II.iii).

Where the pattern of cost overrun is often strikingly similar between projects, the causes of overrun typically differ. For the Channel tunnel, changed safety requirements were a main cause of overrun. For the Great Belt link, environmental concerns and accidents with flooding and a devastating fire made the budget balloon. For the Øresund link, it proved more costly than estimated to carve major new transport infrastructure into densely populated Copenhagen, and so on. *What*, exactly, causes cost overrun in major infrastructure projects is more difficult to predict than the fact *that* cost overrun is likely to haunt projects. But knowledge

Table II.iii *Construction cost development in four privately owned transport infrastructure projects. The table includes private projects for which data were available. Constant prices*

Project	Cost overrun (%)
Channel tunnel, UK, France	80
Third Dartford Crossing, UK	20
Second Severn Crossing, UK	20
Pont de Normandie, France	15

Source: Mette K. Skamris, 'Economic Appraisal of Large-Scale Transport Infrastructure Investments'.

of the latter fact – the 'that' of cost overrun – is the appropriate, necessary and sufficient point of departure for the type of risk analysis and management that we advocate in this book and that is sorely lacking in the planning of most major infrastructure projects today. Therefore, at this stage we are more concerned with the 'that' of cost overrun than with the 'what'.

Conclusion: don't trust cost estimates

On the basis of the evidence presented in this chapter, we conclude that the cost estimates used in public debates, media coverage and decision making for transport infrastructure development are highly, systematically and significantly deceptive. So are the cost–benefit analyses into which cost estimates are routinely fed to calculate the viability and ranking of projects. The misrepresentation of costs is likely to lead to the misallocation of scarce resources, which, in turn, will produce losers among those financing and using infrastructure, be they taxpayers or private investors.

An important policy implication for this highly expensive and highly consequential field of public policy is for the public, politicians, administrators, bankers and media not to trust the cost estimates presented by infrastructure promoters and forecasters. Another important implication is that institutional checks and balances – including financial, professional or even criminal penalties for consistent or foreseeable estimation errors – should be developed to ensure the production of less deceptive cost estimates. In later chapters we will see what such checks and balances might look like.

That large cost overruns are common in major projects does not mean that examples of good practice do not exist regarding cost estimation and management. For transport infrastructure projects, such practice is most common for roads. But even for rail, examples of good practice exist, for instance the construction of the Paris South-East and Paris Atlantique high-speed rail lines (TGV) in France which had only small cost overruns, as had the Toronto Danforth rail extension and the Cologne Metro.[15] For the average major project, however, there is substantial scope for improvement in cost-estimation procedures and in the institutional arrangements to control costs.

3 The demand for megaprojects

Demand forecasts are the basis for socio-economic and environmental appraisal of major infrastructure projects. Furthermore, estimates of the financial viability of projects are heavily dependent on the accuracy of such forecasts. According to the experiences gained with the accuracy of demand forecasting in the transport sector, covering traffic volumes, spatial traffic distribution and distribution between transport modes, there is evidence that demand forecasting – like cost forecasting, and despite all scientific progress in modelling – is a major source of uncertainty and risk in the appraisal of major projects.

Channel tunnel, Great Belt and Øresund

The Channel tunnel opened operations in 1994. Traffic forecasts made at the time of the decision to build the tunnel predicted 15.9 million passengers on the Eurostar trains in the opening year. Actual traffic in 1995, the first full year of operations, was 2.9 million passengers, or 18 per cent of passengers predicted. After more than six years of operations, in 2001, the number of passengers had grown to 6.9 million, or 43 per cent of the original estimate for the opening year (see Figure 3.1). Total passenger traffic (including passengers on shuttle trains, in addition to Eurostar trains) was predicted at 30 million for the opening year; actual total passenger traffic in 1997 was half of this. Rail freight traffic was predicted at 7.2 million gross tonnes for the opening year; actual rail freight traffic was 1.3 million gross tonnes in 1995, or 18 per cent of freight predicted. In 2001 freight traffic had grown to 2.4 million tonnes, that is 33 per cent of the freight traffic investors were told would frequent the tunnel in the opening year. Unfortunately, Eurotunnel has been unable to provide us with data that would allow a similar comparison for shuttle traffic. In sum, more than six years after the tunnel was opened, traffic for which data were available was still less than half of that predicted for the opening year.

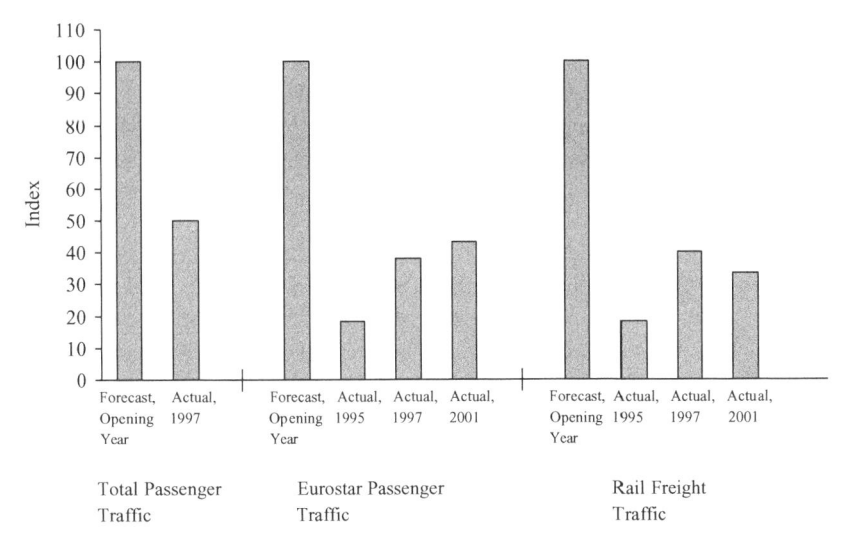

Figure 3.1 Forecast and actual traffic for the Channel tunnel (opening year 1994). Data have been indexed with forecasts for the opening year set at 100

Traffic forecasting for the Channel tunnel has been carried out since 1985. The forecasts are commissioned by Eurotunnel, and while the first forecasts today appear to be optimistic, later predictions, until the years 1989–91, exhibited even more optimistic figures.[1] We explain this by the need to preserve the confidence of the shareholders and of the banks when new cost overruns became apparent in the project. After the tunnel had been completed, the forecasts became more conservative. It should also be noticed that the traffic forecasts commissioned by British Rail and the French SNCF have shown significantly different outcomes. The differences need to be understood in the context of the cautious approach of the British Treasury.

When the construction law for the Great Belt link was ratified by the Danish Parliament in 1987, forecast daily traffic in the opening year was estimated at 19,700 rail passengers and 9,800 road vehicles.[2] In 1997, after the rail link opened on 1 June, actual daily rail passenger traffic was 19,300, in other words almost on target.[3] Over the next two years rail passenger traffic fell, to 17,600 passengers per day in 1999, but only to grow again, to 19,500 per day in 2001.[4] For the road link, which opened on 14 June 1998, actual daily road traffic was 16,990 vehicles or 73 per cent higher than forecast. Since then road traffic has grown every year, to 21,210 vehicles per day in 2001.[5] Actual road traffic on Great Belt thus overshot even the highest forecasts (for the development in forecasts, see Figure 2.2).

For the Øresund link, when the construction law was ratified by the Danish Parliament in 1991, forecast daily traffic in the opening year was estimated in the order of 8,000–10,000 road vehicles with 10,000 vehicles per day as the benchmark, and 16,500–19,000 rail passengers.[6] The link opened to traffic 1 July 2000 and data for the first six months of operations indicated that traffic would be lower than expected.[7] In January 2001 the Danish–Swedish management of the Øresund link therefore agreed to cut prices in an attempt to increase road traffic and revenues, but without much success. In 2001, the first full calendar year of operations, average daily traffic was 8,100 road vehicles and 13,400 rail passengers, both well below the forecasts.

Other transport infrastructure projects

The Channel tunnel is not the only major project that has experienced serious revenue problems. Table III.i compares the Chunnel with selected projects.[8] Below we review more systematically the experience with demand forecasting for a number of countries and studies.

In Germany, the forecasting of the impacts of transport infrastructure projects is based on a comprehensive demand forecasting exercise for the total transport system.[9] Although well founded, this procedure does not guarantee that forecasting failures do not occur, as evidenced by considerable deviations of actual from forecast data. When the forecasts for the German Federal Investment Plan 1985 were presented, the figures of the 'rapid change scenario 2000' for car kilometres travelled were already below the actual data. The development of rail freight transport was not only wrongly predicted by an order of magnitude, even one of the major predicted tendencies was wrong: instead of the predicted upturn, the railways experienced a decline of freight volume transported. The forecasts for the 1992 plan show the same tendency to prediction errors.[10] While the forecasts for road freight transport are significantly below the actual trends, the forecast for rail freight overrun actual figures by more than 100 per cent.

In the UK, the Department of Transport is 'reasonably satisfied' if the original forecast of traffic flow for the year after opening a given section of road is within 20 per cent of actual flow for that year.[11] In a study carried out by the Department, it was found that twenty-two of forty-one road schemes analysed were within this limit. The nineteen schemes with wider variation contained examples where differences between forecast and actual flows ranged from −50 to +105 per cent.[12]

Table III.i *Examples of projects that have experienced serious revenue/benefit problems*

Project	Actual traffic as percentage of forecast traffic, opening year
Calcutta metro, India	5
Channel tunnel, UK, France	18
Miami metro, USA	15
Paris Nord TGV line, France	25
Humber Bridge, UK	25
M65 Huncoat Junction to Burnley Section, UK	35
Tyne and Wear metro, UK	50
Mexico City metro	50
Denver International Airport	55

Source: Mette K. Skamris, 'Economic Appraisal of Large-Scale Transport Infrastructure Investments'. The percentage for Denver International Airport was calculated from Andrew R. Goetz and Joseph S. Szyliowicz, 'Revisiting Transportation Planning and Decision Making Theory: The Case of Denver International Airport', *Transport Research*, part A, vol. 31, no. 4, pp. 263–280.

In another UK study, the National Audit Office identified forty-one Department of Transport and Welsh Office road projects where actual traffic flows were below forecast flows to a degree that the authorities might have adopted lower design standards with possible savings of some £225 million. Similarly, actual traffic flows for twenty-seven projects could have justified higher design standards than those adopted at an additional cost of approximately £160 million.[13] The National Audit Office notes that the authorities did not evaluate the consequences of inaccurate forecasts for individual road schemes in the light of actual traffic flows, nor did they in practice attempt to quantify the costs and benefits actually achieved for such schemes, that is no attempt was made to learn from experience by carrying out *ex post* evaluations of costs and benefits.

Forecasts of rail transport seem to be even more problematic than forecasts of road traffic. The US Department of Transportation study mentioned in the previous chapter found that for virtually every project the divergence between forecast and actual ridership was wider than the entire range of the critical decision variables. Actual ridership was 28 to 85 per cent (average 65 per cent) lower than forecast ridership, meaning that forecasts overshot actual development by 38 to 578 per cent (average 257 per cent).[14]

26   Megaprojects and Risk

In the Transport and Road Research Laboratory study mentioned in Chapter 2, forecast and actual ridership could be compared for nine metro systems. Only in two of these cases was the forecast ridership approximately achieved (actual ridership less than 20 per cent lower than forecast ridership). Of the remaining seven metro systems, actual ridership was 20 to 50 per cent lower than forecast ridership in two systems, 50 to 70 per cent lower for four systems, and 70 to 90 per cent lower for one system. Hence forecasts overshot actual development by an average of more than 100 per cent.[15]

The Aalborg University study mentioned in Chapter 2 contains comparable data for forecast and actual traffic for 210 rail and road projects.[16] Figure 3.2 shows the distribution of inaccuracy of traffic forecasts for all 210 projects in the sample split into rail and road projects. Inaccuracy is measured as actual minus forecast traffic in percentage of forecast traffic. Thus perfect accuracy is indicated by zero; an inaccuracy of −40 per cent, for example, would indicate that actual traffic was 40 per cent lower than forecast traffic, whereas an inaccuracy of +40 per cent would mean that actual traffic was 40 per cent higher than that forecast. The most noticeable attribute of Figure 3.2 is the striking difference between rail and road projects. For rail we make the following observations (again the conclusions are all statistically significant and most likely conservative):

- The average inaccuracy of rail passenger forecasts is −39 per cent (standard deviation, sd = 52). Hence actual traffic was on average 39 per cent lower than forecast traffic, meaning that forecasts were overestimated on the average by 65 per cent;
- For the rail projects, in 85 per cent of the cases the difference between actual and forecast traffic is more than ±20 per cent. Even if we double the threshold value for inaccuracy to ±40 per cent, we find that in 74 per cent of the cases the difference between actual and forecast rail traffic is larger than this;
- There is a massive and highly significant problem with inflated forecasts for rail projects. For two-thirds of the projects, forecasts are overestimated by more than two-thirds;

For road projects we find:

- The average inaccuracy of vehicle traffic forecasts is 9 per cent (sd = 44), namely actual vehicle traffic, was on average 9 per cent higher than forecast traffic;
- For the road projects, in 50 per cent of the cases the difference between actual and forecast traffic is more than ±20 per cent. If we double the threshold value for inaccuracy to ±40 per cent, in 25 per cent of the cases the difference between actual and forecast traffic is larger than this;

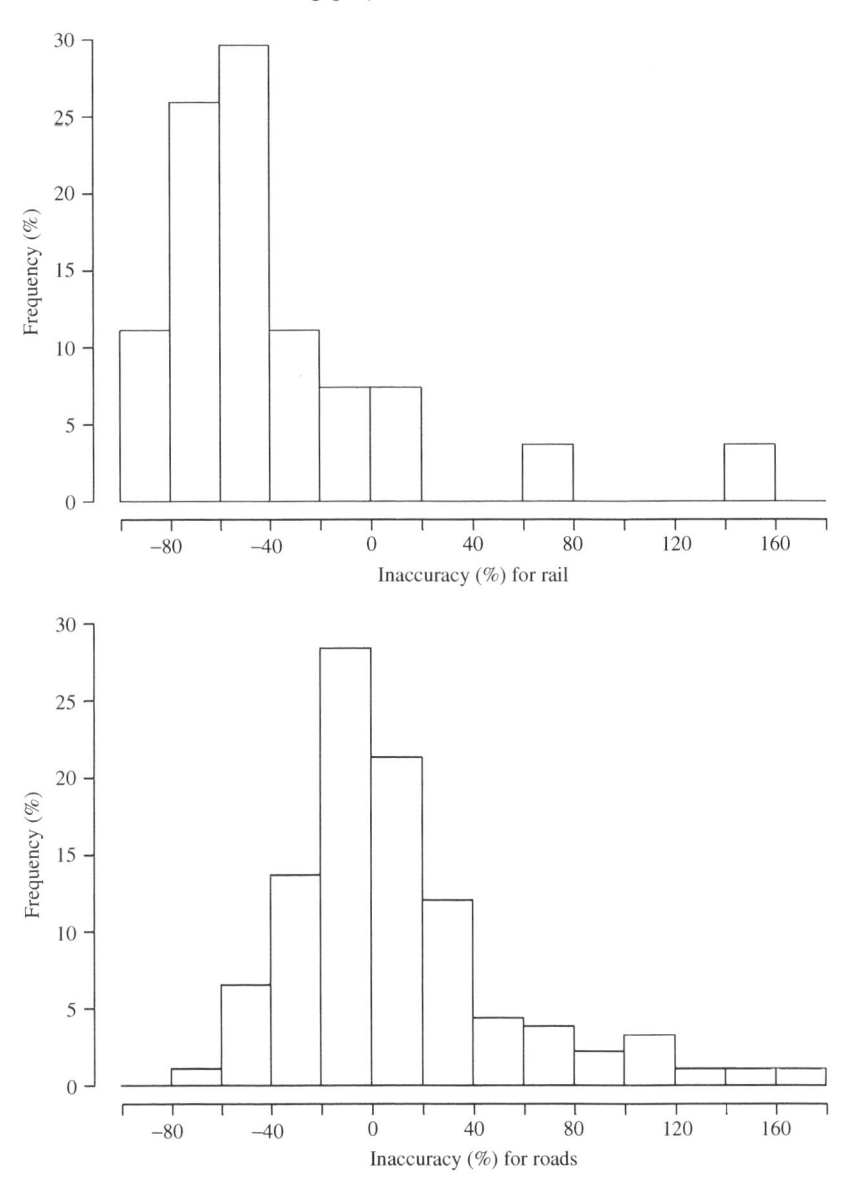

Figure 3.2 Inaccuracies of traffic forecasts in 210 transport infrastructure projects (27 rail, 183 road). Inaccuracy is measured as actual minus forecast traffic in percentage of forecast traffic. The figure shows that rail forecasts are substantially more inaccurate and biased (inflated) than road forecasts

28   Megaprojects and Risk

- There is no significant difference between the occurrence of inflated versus deflated forecasts for road traffic.

Testing the difference between rail and road, we find with a very high level of significance that rail passenger forecasts are more biased (inflated) than road vehicle forecasts. However, there is no indication of a significant difference between the standard deviations for rail and road forecasts; both are high, indicating a large element of uncertainty, and thus risk, for both types of forecast.

We will now turn our attention to the reasons why traffic forecasts are misleading.[17]

Reasons for demand prediction failures

As shown, the data available on the accuracy of traffic forecasts indicate a rather low reliability of professional forecasting. A careful diagnosis of the reasons will provide a better understanding of the multiple factors that contribute to prediction failures.[18] We distinguish seven reasons for prediction failures.

1.   Methodology applied

Most forecasts apply sophisticated multivariate, so-called logit or probit techniques within scenario models. Without neglecting the value of the many scientific debates of the appropriateness of methods and econometric software packages, one can state that applying the wrong method is in general a minor reason for forecasting failures. It has to be recognised, however, that despite all sophistication of econometric methods there are still major weaknesses of modelling which contribute to the high uncertainty of forecasting results. This is particularly true for freight transport modelling. Freight transport patterns change flexibly and are increasingly the result of integrated planning of production and logistics within the single firm. Patterns are, therefore, driven by highly specialised demands that can hardly be aggregated to describe global trends with sufficient accuracy. Technical progress and organisational change within and between companies usually are not continuous processes over time. And it is very hard to predict to what extent firms are able to make use of technological comparative advantages, as has been noticed in the case of the European railways. Although most experts have estimated an improved market potential for this transport mode in the growing international freight market in Europe, the railways have lost market shares for reasons that are hard to predict, such as the inflexibility of the railway companies, the missing harmonisation in the transport markets and the decrease of the relative costs of road transport which has occurred despite talk of allocating the true costs of transport to modes.

2.   *Poor database*

Poor data is a more important reason for prediction failures than methodology. In many countries there is no continuous generation of field data. This means that traffic demand models cannot be calibrated on the basis of observed traffic behaviour (the so-called revealed preference approach). This gap can partly, but not completely, be closed by stated preference analysis (asking people what they would do in defined decision situations). The problem is that actual behaviour of people may, and often does, deviate substantially from the stated preferences.

3.   *Discontinuous behaviour and the influence of complementary factors*

Forecasting analysts are often surprised by the fact that the behaviour of people keeps stable over a time interval although significant changes of influencing factors, such as travel time or cost, have occurred. On the other hand, sudden changes in traffic demand may occur if complementary factors, which cannot be modelled explicitly by demand functions, such as the attractiveness of stations, shopping facilities, Park and Ride, user safety, information and guidance, clean carriages and access paths, adjusted town development and housing policy, and so on, have changed.[19]

4.   *Unexpected changes of exogenous factors*

Sudden changes of exogenous factors can hardly be controlled by demand modelling and scenario techniques.[20] For instance abrupt social and political changes such as the breakdown of the communist regimes and the change in the east–west relationships in Europe are not predictable. Another example is the development of energy prices, which underlies influences that are hard to predict, as for instance in the cases of the two oil crises in 1973 and 1979. The main failure of the German traffic forecasts from 1985 mentioned above was that it assumed a steady increase in real energy prices. But after 1985 real energy prices fell rapidly, therefore the presumed constraining market force for the growth of car traffic and aviation vanished.

30   Megaprojects and Risk

5. *Unexpected political activities or missing realisation of complementary policies*

Unexpected political activities, or unfulfilled promises for political actions, have become a problem since the scenario-technique of forecasting became popular. Usually scenario forecasts are prepared in a way where the political side describes that part of the future world that is influenced directly by political actions. Examples are taxation policy, regulations and complementary activities for the project under investigation (for example access roads, urban/spatial development or international agreements). But stated political preferences and actual political activities are often very different. We find a central example of such differences in the European Union. While the Green and White Papers on the common transport policy promote sustainable development in words, actions that would match the words still lag behind and actual development proceeds in the opposite direction from the established policies. The state of discussion for CO_2 taxation or driving regulations for lorries are cases in point. Consequently, ecologically oriented forecasting scenarios may very well fail for the transport sector, as has happened in both Germany and Denmark.

> Although sophisticated demand models seem objective and hard to manipulate, it is technically easy to tune the models in ways so that 'plausible' or 'desirable' results are achieved.

6. *Implicit appraisal bias of the consultant*

Analysing the direction of prediction errors over time and relating them to the dominating stated preferences of the public, one cannot fail to notice that consultants often integrate political wishes into their forecasting framework. Although sophisticated demand models seem objective and hard to manipulate, it is technically easy to tune the models in ways so that 'plausible' or 'desirable' results are achieved. Another reason for bias produced by means of sophisticated models is their calibration with data from the home countries of the consultants. For example, there were striking differences between the demand forecasts for the high-speed rail link between Melbourne and Sydney, where the Japanese consultants produced very high, the French high and the US consultants low figures, obviously influenced by their home experience.

7. *Appraisal bias of the project promoter*

Estimates of future traffic produced by the project promoter may be even more prone to bias than estimates produced by consultants since the promoter often has an obvious interest in presenting the project in as favourable a light as possible and may be under less pressure than consultants to enforce professional standards.[21]

Taken together these factors explain the poor performance record of megaproject demand forecasting. They also explain why differences between forecast and actual development are not statistically random, but instead biased, as shown above. Such bias, and the interests behind them, may ultimately threaten project viability. Viability is the subject of the next chapter.

> Decision makers are well advised to take with a grain of salt any traffic forecast that does not explicitly take into account the risk of being very wrong. For rail passenger forecasts, and especially for urban rail, a grain of salt may not be enough.

Conclusion: don't trust traffic forecasts, especially for rail

On the basis of the evidence presented above, we conclude that the traffic estimates used in decision making for rail infrastructure development are highly, systematically and significantly misleading. Rail passenger traffic forecasts are consistently and significantly inflated. For road projects the problem of misleading forecasts is less severe and less one-sided than for rail. But even for roads, for half the projects the difference between actual and forecast traffic is more than ± 20 per cent. On this background, decision makers are well advised to take with a grain of salt any traffic forecast that does not explicitly take into account the risk of being very wrong. For rail passenger forecasts, and especially for urban rail, a grain of salt may not be enough.

4 Substance and spin in megaproject economics

The key variables of financial viability for any major project are costs (investment, financing, operations and maintenance) and revenues (mainly tolls in the case of transport projects). For each variable, forecast values may be different from actual values as documented above. A risk therefore exists that actual project viability may be substantially different from forecast viability.

The difference between forecast and actual viability may be so large that if the actual viability had been known for a given project, decision makers might have resolved:
 (i) not to implement the project;
 (ii) to implement the project in another form; or
 (iii) implement another project.

In other words, non-viable projects, or projects that are less viable than forgone projects, may be implemented not because they are viable but because their viability was inaccurately predicted. The result would clearly be an inefficient use of resources.

Channel tunnel, Great Belt and Øresund

For the Channel tunnel, original estimates of viability have been rendered irrelevant by actual developments which have taken the project on a roller-coaster ride from expected high profitability to several near-bankruptcies. Most observers today consider the commercial viability of the Channel tunnel unproved and the prospect uncertain for original investors making a satisfactory profit.[1]

After being issued at £3.50 per share on 9 December 1987, by mid-1989 initial optimism had spiralled Eurotunnel share prices to more than three times this value, above £11.00. Then delays and cost overruns hit the project, resulting in capital shortage and crisis. Shares plummeted, bringing prices below one quarter of the peak value. At the time when demonstration runs went through the tunnel in October 1994, CNN reported that share prices had reached an all-time low.[2] Since then the

Substance and spin in megaproject economics 33

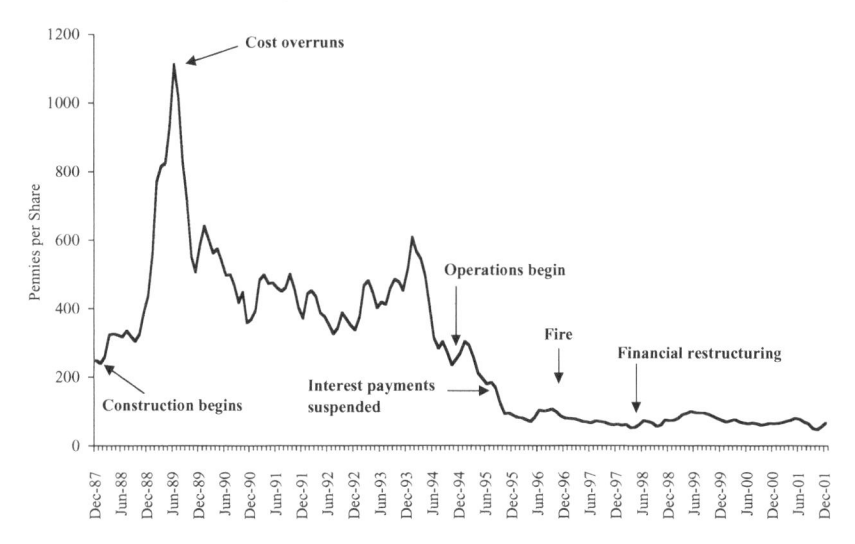

Figure 4.1 Would you invest in this project? That is the basic question as regards viability. The figure shows the roller-coaster ride of Eurotunnel share prices, 1987–2001 (monthly averages)
Source: Data from Datastream, March 2002.

shares have again lost more than two-thirds of their value. Investors' trust in the project has not been regained despite financial restructuring; shares traded at about 65p in 2001, more than six years after the opening to traffic. This is more than 80 per cent below the issue and 95 per cent below the peak value (see Figure 4.1).

After the shocks of massive cost overruns, investors began to worry that Eurotunnel had been as optimistic predicting traffic in the tunnel as it had been predicting costs. Such worries have proven well founded as we saw above when comparing forecast and actual traffic. The result is that until now, the Channel tunnel has not been viable, that is revenues have not been able to cover costs. To avoid bankruptcy, Eurotunnel in September 1995 suspended interest payments on its huge debt. From 1995 to 1998 the company was struggling to arrive at an arrangement for survival with the 225 banks and 750,000 shareholders who have invested in the project. In December 1997 the British and French governments decided to help Eurotunnel by extending the concession to operate the tunnel by thirty-four years to 2086, and in April 1998 an arrangement for financial restructuring was implemented, reducing the company's debt and financial charges, hence giving the company another lease for proving its viability under new and improved circumstances.

According to *The Economist*, however, fares, traffic and market share will all have to increase substantially if Eurotunnel is to break even.[3] Whether this is a realistic scenario only time can tell. Add to this the decision by the European Union to abolish duty-free trade, the delays and cost increases for the construction of the high-speed rail link between the tunnel and London, and the merger of Eurotunnel's two main competitors P&O and Stena ferry lines. Given such policy and market risks, the future viability of Eurotunnel is anybody's guess. Consequently investors continue to worry whether and when they will see a return on their money, and we should not be surprised if more chapters unfold in the high drama of keeping Eurotunnel alive.

For the Great Belt link, being a state-owned enterprise, there are no share prices, and, therefore, no direct market assessment of the viability of the project. Originally the project was set up with separate financial arrangements for the road and rail links. The road link would be owned by Great Belt Ltd and paid for by user tolls with an estimated payback period of initially 12 years. The rail link would be owned by the Danish State Railways, DSB, which would pay back the link over 30 years through an annual fee set to cover construction costs and interest payments. However, the rail link proved non-viable even before rail services started in June 1997.

Cost overruns of 110 per cent on the rail tunnel and of 30 per cent on rail infrastructure on the link (constant prices) inflated the annual fee to be paid by the State Railways to levels that the Railways claimed were prohibitive.[4] At the same time rail revenues were undermined on three fronts. First, a planned three-year head start of rail over road was cut to one year owing to delays on the rail tunnel. As a result, actual rail revenues would be lower than those forecast. Secondly, capacity problems on connecting tracks were identified, also affecting expected revenues. Finally, the State Railways declared that they generally might have been too optimistic in earlier forecasts of rail traffic.

Against this background, the construction law for the Great Belt link was revised in 1996 with a view to solving the viability problems for the rail link. Even though cross-subsidisation is generally considered unsound public policy in both Denmark and the EU, this was the solution settled upon by the Danish Parliament for the Great Belt link. The non-viable rail link would be subsidised with revenues from the road link. Ownership of the rail link was shifted from the Danish State Railways, which consequently never got to own the link as originally planned, to Great Belt Ltd, which reluctantly became owner of both the road and rail links. The State Railways would pay a user fee for running trains across the Belt, a fee substantially lower than that required to pay back the rail link.

At the same time road tolls would be lowered by 20 per cent. Great Belt Ltd would then pay back the total costs of the rail and road links from its combined revenues from the two links.[5]

Before the company was forced to take over the rail link, they expected to be making a profit after approximately thirteen years (the estimated payback period of the road link). After the revised construction law and cross-subsidisation, the payback period was extended to thirty-five to forty years.[6] But after the road link opened, the source of cross-subsidies, namely car traffic, turned out to be larger than forecast. Thus in 2001 the expected payback period had been shortened to 26–9 years. The actual payback period will depend on a number of things, including traffic development, the future level of the State Railways' user fee and, perhaps most importantly, the costs of financing the huge debt on the link.

The first full annual accounts of Great Belt Ltd, for 1999, show dramatically the importance of financial costs to viability. Before financial items are included in the accounts, the results are considerably better than the budget because of higher than expected revenues from the road link of DKK 280 million. After including financial items, however, the result is DKK 150 million below the budget. The main reason is that approximately 10 per cent of the company's debt was in Japanese yen and during 1999 the yen increased in value more than 25 per cent against the euro and the Danish kroner.[7] After this, the Danish Ministry of Finance decided that Great Belt Ltd and other enterprises backed by sovereign guarantees may be exposed only in euros and Danish kroner.[8]

Another main source of financial risk for Great Belt Ltd is changes in the interest rate. The real interest rate realised for the company in 1999 was 4.02 per cent, close to the 4 per cent that is assumed for the company's long-term budget.[9] Given a debt of DKK 38 billion, even a small increase in the interest rate would dramatically increase the financing costs of the company and affect the viability of the project. The owners of both the Great Belt and Øresund links have decided to keep a large part of the debt in variable-rate loans, and so far they have benefited from this decision with real interest rates as low as 3.2 per cent for Great Belt and 2.7 per cent for Øresund in 2001.[10] The rationale behind this decision is that revenues from traffic are dependent on business cycles in the economy where a recession would lead to lower revenues. A recession normally also results in lower interest rates, however, and hence lower costs. Consequently, variable-rate loans may be seen as a hedge against risks to viability stemming from lower revenues, according to this rationale. But the Great Belt and Øresund links have opened operations in an economy where interest rates are historically low. It is an open question, therefore, whether rates will go lower, even in a recession. This question translates into a risk to project viability: if interest rates are about as low as they will get, then keeping a large part of the debt in variable-rate loans is not a protection against risk but the opposite.

For the viability of the Great Belt link, we conclude that given the construction cost overruns of DKK 7.5 billion (1988 prices), Great Belt Ltd and the Danish taxpayers have been plain lucky that the link opened to a booming economy with historically high levels of road traffic – two-thirds more than forecasted – and historically low interest rates. Without such luck – for instance, a recession combined with higher interest rates – there is real risk that the rail link would not have been the only non-viable part of the project, as is now the case; the whole link might have been at risk. Sensitivity analyses carried out by Great Belt Ltd show that a 10 per cent drop in revenues from the 1999 level would result in a five-year longer payback period. An increase in interest rates of just half a percentage point would increase the payback period by three years.[11] In 1997, the Auditor-General of Denmark similarly calculated that with an increase in interest rates of 2 per cent it would be impossible to pay back the project.[12] For the project to stay viable, therefore, the luck and good management skills of the project company will continue to be important. The Great Belt link is a textbook example of the substantial risks involved in megaproject development and of how sensitive viability is to such risks.

Construction of the Øresund link was ratified by the Danish Parliament in 1991. As for the Great Belt link, a condition for ratification was that the link would be 'self-financing', that is revenues from user tolls would cover costs and no public funds would be spent on the project. This condition was explicitly spelled out both in the political agreement about the project between the main parties in the Danish Parliament and in the agreement between Denmark and Sweden.[13]

When the Minister of Transport proposed the Øresund law to the Parliament on 2 May 1991, it was told that in terms of forecast viability, the project would create net revenues of DKK 50 million per year.[14] However, the Auditor-General of Denmark later found that internally in the Ministry, in the months prior to proposing the law to the Parliament, four appraisals of viability had shown the project to be non-viable, that is revenues did not cover costs over a required 30-year payback period.[15] Neither were these appraisals nor was information about them made available to the Parliament when it made its decision regarding Øresund.

In 1994, the Auditor-General of Denmark reviewed the appraisals and found the budget for the Øresund link 'uncertain'. In a sharp criticism, the Auditor-General pointed out that the Parliament should have been better informed about this before making its decision to go ahead with the project. The Auditor-General concluded: 'The significance of this uncertainty is large as even a small budget increase from the economic assumptions of the law proposal will make it impossible to fulfil the requirement of self-financing [the link].'[16]

Since the Auditor-General wrote this, cost overruns have increased to 26 per cent for the coast-to-coast facility and 68 per cent for connecting links on the Danish side (constant prices), while traffic has been lower than that forecast. As mentioned previously, attempts to increase road traffic and revenues by lowering tolls have failed. Integration of the economies of Eastern Denmark and Southern Sweden appear to be substantially more difficult than envisioned by project promoters (see also Chapter 6). In addition, competition from remaining ferries on the sound is stronger than expected. Against this background the Danish Ministry of Transport, Ministry of Finance and Sund & Bælt Holding Ltd decided to carry out a study of the viability problems for the Øresund link.[17] The study, published in 2002, found that during 2001, the first full year of operations, revenues from road traffic were only 40 per cent of those forecast at the time of decision to build.[18] The study also found that most likely the required payback period of 30 years for the coast-to-coast link will not be met. Likewise, for the Danish access links it may be impossible to pay back the investment at all without subsidies from the taxpayer, according to the study. And the Danish taxpayer is already involved, because payback of the Danish access links is highly dependent on a tax arrangement with Great Belt Ltd where losses on the Øresund link are deducted from revenues at Great Belt before taxation. In sum, at the time of writing this book viability of the Øresund link looked unpromising. Viability will be proven, or disproven, only by future development.

The double risk of urban rail

The Aalborg University study of cost and patronage in transport infrastructure projects quoted in Chapters 2 and 3 shows that urban rail projects on average turn out substantially more costly than forecast.[19] For the forty-four urban rail projects in the Aalborg sample for which data on cost overrun were available, average cost overrun was 45 per cent. For 25 per cent of the projects cost overrun was at least 60 per cent; for 75 per cent of projects cost overrun was at least 33 per cent.

At the same time the study shows urban rail achieving considerably fewer passengers than forecast and thus lower revenues. For the twenty-two urban rail projects in the sample for which data on forecast and actual ridership were available, actual ridership was on average 51 per cent lower than forecast. For 25 per cent of these projects actual traffic was at least

Table IV.i *Cost overrun (constant prices) and ridership for twelve urban rail projects*

	Quartiles (25/50/75%)	Average diff. between actual and forecast development (%)	Standard deviation
Costs	28/45/56	40.3	25.3
Ridership	−67/−52/−34	−47.8	25.6

68 per cent lower than forecast; for 75 per cent of the projects actual traffic was at least 40 per cent lower than forecast.

Viability of urban rail is therefore threatened on two fronts, both as regards costs and as regards revenues. Urban rail is doubly risky in economic and financial terms, and the possibilities for financing cost escalations incurred during construction through increased revenues from more passengers during operations often turn out to be limited.

In order to analyse the double risk of urban rail in a more systematic fashion, the Aalborg study identified all urban rail projects for which data were available *both* for the difference between forecast and actual costs *and* for the difference between forecast and actual ridership. The total was 12 projects.

Table IV.i shows the data for the twelve projects. The double risk with both cost escalation and lower-than-forecast ridership is excessively clear for these projects with an average cost escalation of 40.3 per cent combined with an actual ridership that is on average 47.8 per cent lower than forecast.

With only twelve observations, reservations must be made for small numbers. Yet, the numbers are so significant and are supported so distinctly by the larger number of observations in other parts of the analysis that the conclusion stands firm that urban rail projects are high-risk ventures economically and financially because revenue risks augment cost risks and create projects that are risky to the second degree.

German high-speed rail

The history of the MAGLEV rail project between Berlin and Hamburg reveals some basic reasons for the unreliability of forecasts and financial predictions and their effect on viability. When the project was promoted by the industry in 1992, the estimated costs were DM 5.7 billion. The passenger forecast was 15 million per year (for the year 2010). When the Cabinet decided in 1994 to give the green light for the project, these figures had already been corrected to DM 8.9 billion and 14 million passengers, still subject to very optimistic assumptions. On this basis a public–private partnership was established, allocating DM 5.6 billion to the public (for the cost of infrastructure) and DM 3.3 billion to the industry consortium (for the operation investment). The financial risk was implicitly allocated to the state. In 1997 the financial scheme was changed again. The new cost and passenger figures were DM 9.8 billion and 12 million passengers, respectively. The state now took responsibility for infrastructure costs of DM 6.1 billion, the industry for DM 3.7 billion and the operation risk was completely allocated to Deutsche Bahn AG. After a change in the management of Deutsche Bahn in 1999 an internal forecast came out that only 6 to 8 million passengers might be expected, less than half the figure in the early promotion phase. Also, the cost side was out of control from the viewpoint of the Deutsche Bahn management. Against this background the new chair of the holding, a former industrial manager, refused to take any risk into the Deutsche Bahn budget. As the industry was not willing to take on further risk and the same held for the federal government, which had changed by the end of 1998, the project was stopped.

The chronology of the German MAGLEV project clearly demonstrates the political mechanism behind the promotion of megaprojects if the allocation of risk and the performance specifications are not made clear from the beginning. The promoters tried to foster the political process towards a public decision through strategic figures on transport volume and costs, at the same time cultivating the political landscape with lobbying. This was done so efficiently that the Minister ignored all warnings before he in 1994 proposed the project to the Cabinet, which decided in favour of it. The Minister even ignored his own Scientific Advisory Council which provided him with two negative reports on the project. When the promotion process is off to a successful start it is hard to stop again because the promoters and the Ministry work together and produce strategic figures for costs, benefits and viability. It takes outside forces to stop this process, such as private capital or a new federal government unwilling to take on the risks that promoters routinely downplay and then attempt to place on others. Below we will argue that in megaproject development it is crucial that outside checks and balances are institutionalised to restrain and govern a process that otherwise tends to become an anarchic and self-serving means for rent seeking by special-interest groups.[20]

Where the new management of the Deutsche Bahn AG could stop the MAGLEV project early enough to avoid taking on excessive risks, this has not been possible for other German high-speed rail projects. The reason for this failure in risk management is that practically all high-speed rail

projects in Germany have run through the following cycle of evolution:
(1) Rough estimation of benefits, costs and revenues within the standard evaluation procedure of the Ministry of Transport;
(2) Integration of the project in the Federal Investment Plan for Long Distance Transport;
(3) Start of detailed design of the project, check for spatial integration by the states (*Länder*);
(4) New requirements set by the *Länder* as preconditions to implementation;
(5) New cost estimation carried out by a planning agency, control activity by the Federal Railway Agency, revisions of forecasts, agreement between the Federal State and the Deutsche Bahn AG on cost sharing, establishment of financial plans and allocation of public payments to the future fiscal budgets;
(6) Final design of the project, negotiations with communities, treatment of objections of citizens, expropriation processes, new requirements set by the communities;
(7) Construction of the project, reporting of actual cost development to a small group of officials from the involved agencies and the ministries of transport and finance as well as the Deutsche Bahn AG.

The high-speed rail link between Cologne and Frankfurt is a case in point. It was originally designed with one stop between the two cities and a total cost of DM 5.4 billion. After step (5) in the process above, one stop had turned into five and as many new stations. Consequently, average speed would go down as would demand, from an estimated 22 to 18 million passengers per year. Costs were now re-estimated at DM 7.8 billion. This, then, was the basis for the cost-sharing agreement between the Federal Government and Deutsche Bahn AG. According to present estimates, costs will amount to DM 10 billion, or almost twice the budget originally planned, and there will be fewer passengers.

In the case of the high-speed rail link between Nuremberg and Munich, Deutsche Bahn AG favoured a route via Augsburg, because existing tracks could be used. The state of Bavaria, however, favoured a route via Ingolstadt. This was estimated in stage (5) of the process to cost DM 4.7 billion, which caused heavy criticism by the Auditor-General. The Federal Railway Agency (EBA), which is the public control institution for the federal railways in Germany, then arbitrarily cut the cost estimate by DM 0.9 billion, or round about the cost difference between the two project variants. Following this, the former management of Deutsche Bahn AG was placed under extreme pressure by the Federal and Bavarian governments to agree with a financial plan based on the reduced budget of DM 3.8 billion. According to recent estimates, costs will reach at least DM 5.4 billion.

The experience is similar for other German rail projects. Taken together, a risk total of about DM 6 billion has been allocated to the Deutsche Bahn AG in the course of high-speed investment planning. Before the railway reform in 1994, when Deutsche Bahn was a public institution, such a problem would have been solved by muddling through via medium-term government budgets. But since the reform, when Deutsche Bahn became a stockholding company under private law with its stocks fully owned by the federal state, risks have had to be managed privately by the new company, Deutsche Bahn AG, which accordingly has fallen into deep financial crisis. The company has resorted to spending parts of the budget planned for rehabilitation work of the existing rail network to finance the construction work for the few, new large-scale projects. As a consequence, currently there are 2,500 low-speed sections in the railway network, causing continuous delays and disturbances. The attractiveness of the railway services to the users has suffered, which in turn has put the balances of the company in a downward spiral.

These examples show why, in Germany, mega transport infrastructure projects typically turn out to be twice as costly as originally planned and often with lower incomes: manifold political interests are nested and hidden in the process after the general approval of a federal infrastructure investment plan by Parliament. Key actors are interested in increasing performance requirements regardless of costs as well as in keeping the transparency of this cumulative process as low as possible. When carrying out the research for this book, for instance, it was not possible to get those involved to reveal the responsibilities for the decision steps taken in the German projects, not even through contacting members of the board of the Deutsche Bahn Netz AG, the infrastructure company of the German Railways. The system is closed in on itself and thus is presented as a case study for a problem that we will return to in Chapter 8: lack of transparency, which is often related to high risk through lack of accountability.

Compared to the German projects, the Channel tunnel, Great Belt and Øresund links look almost like success stories, especially for the later stages of project development. Fortunately, the German railway reform has made it more difficult to play the game of hiding (and passing on) risks, costs and viability as described above; eventually the results will show up in the balances of the privatised railway company.

Other transport infrastructure projects

The pattern found for the Channel tunnel and Øresund link, and for the German high-speed rail projects, with over-optimistic estimates of project viability in the initial phases of planning, is also found for other projects.

Table IV.ii *Examples of projects that have experienced viability problems*

Project	Construction cost overrun (%)	Actual traffic as percentage of forecast traffic, opening year
Humber bridge, UK	175	25
Channel tunnel, UK, France	80	18
Baltimore metro, USA	60	40
Tyne and Wear metro, UK	55	50
Portland metro, USA	55	45
Buffalo metro, USA	50	30
Miami metro, USA	35	15
Paris Nord TGV, France	25	25

Source: Mette K. Skamris, 'Economic Appraisal of Large-Scale Transport Infrastructure Investments'.

Table IV.ii compares the Channel tunnel to a number of other projects that have also experienced viability problems.

The study by the Auditor-General of Sweden of fifteen road and rail projects referred to earlier found that projects 'live on' in the planning process despite major declines in viability: if early estimates of viability indicated that a project was viable and thereby placed it on the planning agenda, the project tended to stay on the agenda no matter how viability developed, jeopardising the effective use of public funds.[21] The Auditor-General of Sweden further observed that in the majority of instances of cost changes in projects, the responsible authorities did not recalculate project viability. Recalculation of viability was done only in six instances for four road projects. These projects ended up with returns 8 to 40 per cent lower than originally estimated. One SEK 791 million road project ended up with a negative cost–benefit ratio after going through three rounds of recalculations. This project, like the others, was implemented.[22]

The US Department of Transportation study of ten rail transit projects calculated viability by cost-effectiveness analysis, which related cost to ridership.[23] As mentioned, cost overruns in the ten projects ranged from –10 to +106 per cent, whereas actual ridership was 28 to 85 per cent lower than forecast ridership. The result was actual costs per passenger on the average 500 per cent higher than forecast costs (ranging from 190 to 870 per cent) and, accordingly, an actual project viability much inferior to that projected.[24] For the ten US cases, the divergence between forecast and actual cost per passenger was often larger than the entire range of values for this measure with respect to all the alternatives from which a project was selected. The study concludes, 'it is certainly

Substance and spin in megaproject economics   43

possible that decision makers acting on more accurate forecasts of costs and future ridership for the projects reviewed here would have selected projects other than those reviewed here'.[25]

The results from the Transport and Road Research Laboratory study of twenty-one rail transit systems in developing and newly industrialised nations show a pattern similar to that of the US study: costs were underestimated by 20 to 100 per cent and ridership was overestimated by 25 to 225 per cent in the majority of cases, resulting in actual viability being substantially lower than forecast viability.[26] Only three out of the eleven transit systems for which costs and revenues could be calculated showed a surplus (Hong Kong, Seoul, Singapore) despite the fact that during planning of the transit systems in question most governments were led to believe that the systems would be financially viable.[27] One explanation of the difference between actual and forecast viability identified by the authors of the study was 'over-optimism in the planning phases'.[28] Surely the last thing developing nations need is projects that detract from economic development through negative viability.

> The World Bank has called for not only more accuracy in estimates of viability, but also more honesty.

Other infrastructure investments

Data for other types of project than transport infrastructure show that the pattern is general: over-optimistic forecasts of viability are the rule for major investments, rather than the exception.[29] A review of the World Bank's overall project portfolio, the so-called Wapenhans report, documented an increasing number of poorly performing infrastructure projects.[30] A main cause was over-optimistic estimates of project viability. Another was inadequate attention to risk and uncertainty. Other World Bank studies of large numbers of projects have found no indication that the problem is diminishing. On the contrary, a widening gap between forecast and actual viability has been documented for these projects.[31]

World Bank projects are typically subject to more careful appraisal and control than most other infrastructure projects, but even for infrastructure projects undergoing the Bank's relatively rigorous procedures, consistent patterns of inflated project viability have been found.[32] Against this background, the World Bank has called for not only more accuracy in estimates of viability, but also more honesty.[33] We agree that honesty is desirable, and we find that honesty is best improved through improved accountability. In what follows, we will focus on increased accountability as a means to improve project appraisal and decision-making processes for large infrastructure investments.

Trying to understand why viability is so consistently overestimated across major projects, two researchers from the Massachusetts Institute of Technology found that a basic problem in this context is that the incentives to produce optimistic estimates of viability are very strong and the disincentives weak. Accountability is low, and politicians who underestimate costs in order to have projects approved are rarely in office when actual viability can be calculated, if it ever is. Contractors and others with special interests in major projects are also eager to have their proposals accepted. Because contractual penalties for producing overoptimistic tenders are often relatively small, major project costs are often underestimated even during tendering.[34] Similar results have been found in research conducted at the University of California, Los Angeles.[35]

Uncertainty in estimating viability is related in this way not only to the innate difficulty of predicting the future but also to power and interests. If the authorities responsible for project development and decision making fail to take this into account, namely if they do not develop the necessary institutional checks and balances for project appraisal, the risk is that the wrong projects are implemented.

- Cost overruns of 50 per cent to 100 per cent in real terms are common, and overruns above 100 per cent are not uncommon;
- Demand forecasts that are wrong by 20 per cent to 70 per cent compared with actual development are common;
- The key problem is lack of accountability, not lack of technical skills or data.

The problem of inflated viability

The main lessons to be learnt from the projects reviewed in this chapter are:
(1) Cost overruns of 50 per cent to 100 per cent in real terms are common for large transport infrastructure projects, and overruns above 100 per cent are not uncommon. Similar figures hold for other types of project, not only transport projects;
(2) Demand forecasts that are wrong by 20 per cent to 70 per cent compared with actual development are common for large transport infrastructure projects. Forecasts for rail appear to be particularly prone to

large overestimates of traffic, often beyond 100 per cent in the cases for which data are available;
(3) Forecasts of project viability for large transport infrastructure projects are often over optimistic to a degree where such forecasts correspond poorly with actual development.

In sum, the result is that there are many more underperforming projects than can be explained by chance alone. The differences between forecast and actual costs, revenues and viability, which have been documented in this and the previous two chapters, cannot be explained primarily by the innate difficulty of predicting the future. The differences are too consistent and too one-sided for this. Instead, the differences may be explained by project proponents succeeding in manipulating forecasts in ways that make decisions to go ahead with projects more likely than decisions to stop them. The key problem here is a lack of accountability for the parties involved in project development and implementation, that is the key problem is not lack of technical skills or poor data:

(1) Because of the time frames that apply to major project development and implementation, politicians involved in producing over-optimistic forecasts of project viability in order to have projects approved are often not in office when actual viability can be calculated;
(2) Special interest groups can promote projects at no cost or risk to themselves. Others will be financing the projects, and often taxpayers' money is behind them, including in the form of sovereign guarantees. This encourages rent-seeking behaviour for special interest groups;
(3) Contractors, who are an interest group in its own right, are eager to have their proposals accepted during tendering. Contractual penalties for producing over-optimistic tenders are often low compared to the potential profits involved. Therefore, costs and risks are also often underestimated in tenders. The result is that real costs and real risks do not surface until construction is well under way.

Needless to say, this state of affairs does not mean that viable projects and projects showing 'good practice' regarding estimated and actual viability do not exist. But there are many more projects with underestimated costs and overestimated benefits than there are projects with correctly estimated costs and benefits, not to speak of projects with overestimated costs and underestimated benefits, which are even more rare.[36] Even less does it mean that standard practices cannot be substantially improved in project appraisal. It does mean, however, that a pervasive and consistent problem exists for both major transport and other infrastructure projects regarding the reliability of viability estimates and that this problem may turn out to be unsolvable by technical means. The problem appears to be related to issues of power and to require countervailing power and institutional change for its solution.[37]

Is delusion necessary to get projects started?

A question left unaddressed above is whether any megaproject would ever be undertaken if some form of delusion were not involved, that is, would projects be undertaken if the true costs and benefits were known beforehand? The data presented above suggest that project promoters and forecasters would answer this question in the negative; they appear to think delusion is necessary to get projects started and they effectively produce deceptive forecasts. We disagree, however, that delusion is generally necessary to undertake projects, and we want to point to the quagmire of economic, legal and ethical issues that follow from practices of delusion.

It is easy to find motives for producing deceptive forecasts of costs and benefits. Politicians may have a 'monument complex', engineers like to build things and local officials sometimes have the mentality of empire-builders. In addition, when a project goes forward, it creates work for engineers and construction firms, and many stakeholders make money. If these stakeholders are involved in, or indirectly influence, the forecasting process, then this may influence outcomes in ways that make it more likely that a project would be built. Underestimating costs and overestimating benefits would be economically rational for such stakeholders because it would increase the likelihood of revenues and profits. Economic and political self-interest also exists at the level of cities and states. Here, too, it may explain cost underestimation and benefit overestimation. Don Pickrell pointed out that transit capital investment projects in the US compete for discretionary grants from a limited federal budget each year.[38] This creates an incentive for cities to make their projects look better on paper in terms of costs and benefits, or else some other city may get the money. Finally, in some quarters underestimating costs is seen as a means to keeping costs low.[39]

Whereas motives are easy to find, it is harder to find project promoters and forecasters who are willing to talk about actual instances of cost underestimation and benefit overestimation. For legal, economic, ethical and other reasons, if promoters and forecasters have intentionally fabricated a deceptive cost or benefit estimate for a project in order to get it started they are unlikely to tell researchers or others that this is the case.[40] We are aware of only one study that actually succeeded in getting those involved in underestimating costs and overestimating benefits to talk about such issues.[41] Martin Wachs interviewed public officials, consultants and planners who had been involved in transit planning cases in the US. He found that a pattern of highly misleading forecasts of costs and patronage could not be explained by technical issues and were best explained by lying. In case after case, planners, engineers and economists told Wachs that they had had to 'cook' forecasts in order to produce numbers that would satisfy their superiors and get projects started, whether or not the numbers could be justified on technical grounds.[42]

One typical planner admitted that he had repeatedly adjusted the cost figures for a certain project downward and the patronage figures upward to satisfy a local elected official who wanted to maximise the chances of getting the project in question started. Wachs's work is unusually penetrating for a work on forecasting. But it is small-N research without statistical significance and Wachs acknowledges that most of his evidence is circumstantial.[43] The evidence does not allow conclusions regarding the project population. Nevertheless, based on the strong pattern of misrepresentation and lying found in his case studies, Wachs goes on to hypothesise that the type of abuse he has uncovered is 'nearly universal' and that it takes place not only in transit planning but also in other sectors of the economy where forecasting routinely plays an important role in policy debates, for instance energy, environmental planning and housing.[44]

Our larger sample of data, which allows statistically valid conclusions, gives support to Wachs's claim: the pattern of underestimated costs and overestimated benefits is found not only in the small sample of projects Wachs studied. The pattern is statistically significant and holds for the project population mean, in other words, for the majority of transport infrastructure projects. The use of deception and lying as tactics aimed at getting projects started appears to best explain why costs are highly and systematically underestimated and benefits overestimated in transport infrastructure projects.[45]

It is, undoubtedly, common for project promoters and forecasters to believe their projects will benefit society and that, therefore, they are justified in 'cooking' costs and benefits to get projects built. Such reasoning is deeply flawed, however. Underestimating costs and overestimating benefits for a given project leads to a falsely high ratio between benefits and costs for that project, which in turn leads to two problems. First, the project may be started despite the fact it is not economically viable. Or, second, it may be started instead of another project that would have shown itself to yield higher returns than the project started, had the actual costs and benefits of both projects been known. Both cases result in the inefficient use of resources and, for public projects, in waste of taxpayers' money. Thus, for reasons of economic efficiency alone the argument that cost underestimation and benefit overestimation are justified must be rejected.

But the argument must also be rejected for legal and ethical reasons. In most democracies, for project promoters and forecasters to deliberately misinform legislators, administrators, bankers, the public and the media would not only be considered unethical but in some instances also illegal, for instance where civil servants would intentionally misinform Cabinet members, or Cabinet members would intentionally misinform the parliament. There is a formal 'obligation to truth' built into most democratic constitutions on this point as a means for enforcing accountability. This obligation would be violated by deliberate misrepresentation of costs and benefits, whatever the reasons for such misrepresentation may be. Not only would economic efficiency suffer but also democracy and democratic accountability.

Our first answer to the introductory question of whether any megaproject would ever be undertaken if some form of delusion was not involved is, therefore, that even if delusion, in the shape of manipulated cost and benefit estimates, were necessary in order to undertake projects, such delusion would not be defensible in liberal democracies for economic, legal and ethical reasons; economic efficiency and democratic accountability would suffer. Our second answer is that delusion is not necessary to undertake projects, because many projects exist with sufficiently high benefits and low enough costs to justify building them. The problem is not that projects worth undertaking do not exist. The problem is that the dubious and widespread practices for underestimating costs and overestimating benefits used by project promoters and forecasters to promote *their* favourite project create a distorted hall-of-mirrors in which it is exceedingly difficult to decide which projects deserve undertaking and which not. The result is, as we observed in the introductory chapter and as even the industry's own organ, the Major Projects Association, acknowledges, that 'too many projects proceed that should not have done'. We would add that projects also exist that do not proceed but should have, had they not lost out, not to better projects but to projects with 'better' delusion, that is 'better' manipulated estimates of costs and benefits.

5 Environmental impacts and risks

We live in a time when the ability for constant learning is considered crucial to the welfare of individuals, organisations and nations. This is the age of the 'learning society'. However, in environmental impact assessment (EIA), which is the main methodology used by decision makers to predict environmental effects of megaprojects, surprisingly little learning is taking place. Or, to put the matter more positively, learning is only now beginning. This is true for megaprojects as well as for other types of project. The reason for the lack of learning is that projects and their environmental impacts are rarely audited *ex post*, and without post-auditing learning is impossible.

A recent study describes this situation as an unfortunate 'stalemate' and concludes: 'there is much scope for raising the profile of post-auditing in EIA world-wide'.[1] One consequence of the current state of affairs is a lack of knowledge about the actual environmental risks involved in infrastructure and other development. Although many studies on environmental impact analysis have been elaborated, and many environmental impact statements made, there is still a general sense among professionals and decision makers that the state of the art is not satisfactory. In what follows we will elaborate on three reasons for the presumed deficiencies of environmental impact assessment:

(i) a lack of accuracy in impact predictions;
(ii) the narrow scope of impacts and their time horizon; and
(iii) an inadequate organisation, scheduling and institutional integration of the environmental impact assessment process in the overall decision-making process.

Environmental impact assessment (EIA)

Contrasting the ambitious general goals of the Brundtland report and of the many national and international papers that exist on global environmental issues, the concrete environmental impact assessments of projects are often restricted to checklisting procedures that stress well-established knowledge on local impacts, while ignoring interregional, global, systemic or long-term effects.[2] As the framework for the performance of environmental impact assessment is still very general and no binding guidelines exist, there is a high degree of freedom for government bodies and consultants in carrying out the tasks involved. In Denmark, improvements in the framework for assessments have taken place with a 1993 circular from the Prime Minister's Office and with the publication by the Ministry of Environment of advice regarding environmental impact assessment of law and other government proposals.[3] Still, these improvements do not have the status of binding guidelines. Regional and environmental groups often complain that their concerns have not been appropriately analysed in environmental impact assessments because of too narrow a scope. Cumulative and indirect environmental effects, which are due to synergism – caused, for example, by the accumulation of impacts across projects – are rarely discussed in environmental impact statements.[4]

In the conventional approach to decision making on major transport infrastructure projects, the process begins with investigating technical solutions and ends with assessment and appraisal (see Chapter 8). Often environmental assessment and risk analysis are the final step in preparing the decision for the government. This holds in particular for the German evaluation approach (*Bundesverkehrswegeplan*),[5] and it has also been identified as a weakness of Danish environmental impact assessment for large projects.[6] The consequences are that environmental impact assessment is often reduced to an instrument of the final design of a project, and/or that the results of environmental impact assessment are not easily accepted by the public and give rise to conflicts.

It is becoming increasingly clear that a better way to organise the process is to start with the environmental impact assessment as early as possible and to include participation of stakeholder groups and citizens from the beginning. A major condition for an environmental impact assessment to be successful is therefore to design it as a feedback mechanism.[7] Findings in a number of countries confirm that there is a need for a new scheme of research, planning and decision making.[8]

EIA in the Channel tunnel and Øresund projects

Analyses exist on the procedures of environmental impact assessment that are helpful for arriving at conclusions for good practice. Below, the environmental impact assessments prepared for the Channel tunnel and Øresund link are used to illustrate this point.

On the UK side of the Channel tunnel, an assessment was originally made of the impacts arising from:

- The two rail tunnels and the third service tunnel;
- The tunnel portal and the roll-on-roll-off terminal for road vehicles at Cheriton;
- The tunnel services complex at Shakespeare Cliff;
- Road freight facilities southeast of Ashford;
- Rail freight facilities north of Saltwood.

The following principal impacts were considered and assessed according to EU Directive 85/337:

- Impacts of changes in the patterns of road and rail traffic movement in the region;
- Impacts of further development stimulated by the operation of the Channel tunnel;
- Site-specific impacts of individual facilities.[9]

No major environmental risks were identified as a result of the assessment. A further investigation of environmental impacts regarding safety, noise and air pollution resulted in an overall positive figure. The authors conclude, 'this report demonstrates that the Channel tunnel represents a unique opportunity to bring about the kind of benefits which people in the environmental movement have long hoped for'.[10] Despite this optimistic view, it is well known that there have been many objections on environmental grounds over the siting of a route for the high-speed link on the UK side to integrate the Channel tunnel with the national rail network. Also, it has been stressed that to balance the environmental damage of new construction in Southeast England, long-distance road traffic should be attracted to the railway.[11] If such comparative benefits are taken into account this might justify some form of public subsidies or support, as has been the case for the high-speed rail link between the tunnel and London.

The fixed link between Copenhagen and Malmö across Øresund was assessed with respect to a comprehensive list of impacts.[12] Impacts on the local and regional environment in the Baltic Sea were included in the assessment. Furthermore, a non-technical assessment covered visual impacts of alternative physical constructs for the link, viewed from sea and from the shore, as well as visual impacts for passengers in cars and trains. Three issues were highlighted in the public debate:

(1) Possible impacts on the salt balance of the Baltic Sea. Even diminutive changes in this balance were seen as possible causes of severe changes in the flora and fauna of the Sea;
(2) Possible effects on the wildlife on the island of Saltholm and its coastal waters, which are breeding grounds for many sea birds;
(3) Increased noise and air pollution from the intensified flow of road traffic caused by the fixed link.

While experts agreed on the assessment of (2), there was substantial dissent with respect to (1) and (3). With regard to (1), experts expressed the judgement that the marine impacts had not been assessed in a comprehensive way, namely together with the Great Belt and Fehmarn Belt links, both of which may also affect the salt balance of the Baltic Sea. For (3), the conclusion of the environmental impact statement that emissions would decrease with a fixed link was challenged. Finally, dissatisfaction was expressed about the procedure for selecting the competent institution that would prepare the environmental impact assessment.

According to Andersson *et al.*, the experience with major transport projects in Denmark shows that environmental impact assessments have been weak in the case of projects adopted by a specific public works act, as in the case of the Øresund link.[13] The researchers propose that the environmental impact assessment should be strengthened by prescribing a scoping procedure in the very early stages of a project, after a notice of intent. A scoping procedure – as it is known in, for example, Canada – makes it possible for stakeholder groups to participate in the design of the impact assessment programme.

The UK experience

Environmental impact assessment has been required for certain proposed road developments in the UK since EU Directive 85/337 was implemented in 1988. The extents to which the requirements of the directive are met with respect to ecological issues have been explored in a review from 1993 of road statements[14] and in another review from 1997 of UK environmental impact statements in general.[15] Main shortcomings identified by these reviews were:[16]

- Widespread failure to indicate the actual size of the proposed development, whether in terms of length or land take for major wildlife habitats, making it impossible to estimate potential ecological effects;
- Widespread failure to provide the data necessary to predict ecological effects;
- Failure in some cases either to carry out appropriate ecological surveys or to report their results;
- Emphasis on vague and descriptive predictions. Only 9 per cent of environmental impact statements in the first (1993) review quantified potential impacts;
- Lack of baseline data;
- Lack of commitment to follow up monitoring;
- Failure to evaluate impacts in terms of local, regional, national or international status of habitats and species;

- Recommendation of mitigation measures that did not address the impacts identified;
- Failure to give detailed prescriptions for proposed mitigation measures or to indicate their likely success.

A review from 2000 of more environmental impact statements and an examination of recent changes in the UK political context for environmental impact assessment of proposed road developments suggested that the ecological assessment of such developments had improved in some respects.[17] For instance, two out of forty impact statements under review now included a commitment to monitoring some aspect of the scheme, whereas no impact statements in either of the previous two reviews included such a commitment. Furthermore, monitoring was discussed as a possibility for the future in four of the forty statements. Despite such signs of improvement the overall impression from the 2000 review was the persistence of many of the shortcomings identified in the two earlier reviews.

The German experience

In Germany the environmental impacts of the federal government's transport projects are analysed in two steps. First, effects that can be quantified and monetarised according to the state of the art are introduced in the Ministry of Transport's cost–benefit analysis. Here the impacts of noise and some impacts of air pollution (expressed in CO-equivalents) are assessed using standardised approaches for quantification and economic evaluation. The result of this first environmental assessment is aggregated with the other costs and benefits stemming from changes of costs of operation, infrastructure, time, accidents and regional benefits. The aggregate outcome is a cost–benefit ratio which is the base for the ranking of a project in the federal transport investment programme (high priority, low priority or rejected). The set of selected high-priority projects forms the investment programme for a period of twenty years.

If a project has passed this first threshold and is considered for the investment programme, a second environmental check becomes obligatory under specified circumstances. If the candidate link is longer than 10 km or if there is an identified risk that biological entities could be disturbed, then a detailed environmental risk analysis is obligatory. This assessment concentrates in the first instance on local impacts. Although no directive exists for the scope of the analysis and the approach to the assessment and evaluation, the studies are in general carried out by qualified consultants according to standards that have developed experientially. This means that a detailed list of criteria is established for the project impacts, and the environmental risk is assessed for every segment using a ranking scale. If a high risk is identified for segments of a link, this may lead to changes in the project design.

Although in principle it would be possible to reject a project because of its high environmental risk, this has not yet happened. But environmental problems identified in the detailed environmental impact assessments have in many cases led to the redesign of projects and, as a consequence, to considerable increases in project costs. Such cost increases have not been considered in the cost–benefit analysis, however, thus violating the logic of the ranking procedure. In this context, it is important to know that according to the German planning set-up, which on the local level is governed by the States (*Länder*), an expropriation of property owners is only possible if: (i) the project is necessary to increase overall social wealth; and (ii) there is no better alternative. Environmental impact analysis has often made it possible for stakeholder groups to show that alternative solutions with lower environmental impacts are feasible compared with the project design promoted by the administration. The interdependencies of environmental impact assessment, activities of environmental groups and the legal process have resulted in: (i) a prolongation of the implementation time for projects – in the case of the first high-speed railway links the elapsed time for the planning process and construction was about twenty years; and (ii) tremendous cost overruns. Here as elsewhere environmental risk often translates into financial risk.

In the case of the high-speed Hannover–Berlin rail link, for instance, it was discovered that the projected alignment of the track would touch the breeding area of rare bustards at some distance. Therefore, to provide the conditions for the project to pass the legal process in due time and in order to guarantee that no damage would occur to the birds from the trains, for a long time a design putting the track underground for some kilometres was considered. The consequence would have been a considerable increase of the investment budget such that the cost for preserving the life of one bustard was calculated at a sum of about DM 33 million. For comparison, the value of a human life is calculated at DM 1.3 million in the cost–benefit analysis used for standard project appraisal. The solution finally settled upon was to change the alignment of the track and to construct dams, which would protect the flight paths of the birds. This solution is much cheaper in construction costs but will result in an increase in travel time of some minutes, which might have impacts on travel demand. This example illustrates the way in which environmental issues tend to dominate the final phase of project approval in Germany with consequent delays, large cost overruns and potentially suboptimal solutions for the infrastructure users.

Facing these undesirable side-effects of the evaluation method and its interrelationships with the planning process, several groups such as the Scientific Advisory Board of the Ministry of Transport, an expert group of the Ministry of Urban and Spatial Planning as well as the Environmental Agency have suggested starting the assessment process with a systems analysis based on clearly defined performance specifications for spatial and environmental policy issues.[18] This could help filter out projects that might effect heavy conflicts with spatial or environmental goals or to redesign them accordingly and include their full environmental costs in the following assessment step of project-based cost–benefit analysis.

Such a procedure would be in line with the idea of strategic environmental assessment (SEA), which is promoted by the EU Commission and is obligatory for future assessments of the Trans-European Transport Network (TTEN) or corridors.[19] The guidelines for TTEN from 1996 explicated the necessity of strategic environmental assessment. In 2001 a Directive of the Commission was adopted (2001/42/EC), which is being applied for the TTENs and will be extended to the whole transport system by 2004.

> Buckley found that the average accuracy of quantified, critical, testable predictions in environmental impact statements was only 44 per cent.

Predicted versus actual outcomes

As pointed out by several observers, one of the major reasons for auditing environmental impact predictions is that it presents the opportunity for environmental impact assessment to learn from experience.[20] But to date there has been a shortage of studies that seek to advance the development of environmental audits and audit methodology. Without auditing to support predictive environmental impact assessments the quality and accuracy of information supplied to the decision-making process will remain largely unproven, and this is the situation today. Compared to the huge number of environmental impact assessments that exist it is remarkable how few studies have been carried out that compare predicted environmental impacts with actual outcomes, and only one of these is of major transport infrastructure projects. The limited number of studies found in the literature have tended to be carried out externally to the decision-making process, by academic researchers, rather than internally, by the responsible authorities and developers themselves.[21]

The studies of Buckley[22] and Culhane[23] stand out in this context.[24] In the first national analysis of the accuracy of environmental impact predictions for any country, Buckley found that the average accuracy of quantified, critical, testable predictions in environmental impact statements in Australia until 1991 was only 44 per cent.[25] Primary predictions, such as those referring to the characteristics of emissions to air and water, came out more accurate on average than secondary predictions (impacts of emissions), such as those referring to ambient air and water quality. The accuracy of individual predictions differed by over three orders of magnitude, with actual impacts ranging from $0.05\times$ to $200\times$ the predicted value.

Culhane found that most of the predictions in the reviewed environmental impact statements were very imprecise, with less than 25 per cent being quantified. However, only a few unanticipated impacts were identified. In sum, only 30 per cent of the impacts were unqualifiedly close to their forecasts, with almost as many rated accurate principally by virtue of the vagueness of the forecast.

A more recent study by Christopher Wood, Ben Dipper and Carys Jones found more reasonable levels of accuracy, but still found that 44 per cent of predictions were not auditable and that the more significant impacts were frequently inaccurate. Finally, very little monitoring of environmental impacts was actually occurring.[26]

In the case of major transport projects, we are not aware of published results from any comprehensive environmental audit studies comparing actual and predicted outcomes for a large number of projects. This is an interesting empirical fact in its own right and it indicates that environmental audits are a rare and new phenomenon for this type of project. So far, only the most advanced projects, such as the Great Belt and Øresund links, employ post-auditing. In the future we are likely to see more such audits from more projects. For the time being, we are aware of only one study that attempts to compare predicted and actual environmental outcomes for major transport projects.[27] This study examines environmental performance in five projects: the Channel tunnel, the Great Belt and Øresund links, the Second Severn bridge between England and Wales and the London Underground Jubilee Line. The study shows that actual environmental impacts during construction often differ significantly from those predicted, whereas differences tend to be smaller after construction is completed and operations begun. It is difficult to compare this study directly with those mentioned above because the sample is too small with only five projects and because several of the projects had not been fully completed at the time of the study. More comparative research is needed on predicted versus actual environmental outcomes in the field of major transport infrastructure projects.

In sum, existing data indicate that there is substantial risk involved in using state-of-the-art predictions of environmental outcomes in decision making on major projects. Such predictions are simply unreliable. Risk analysis and risk management are therefore required as much for environmental issues as for issues of costs, revenues and viability. For many types of major project, including large-scale transport infrastructure projects, empirical data are currently insufficient to generate figures on the means and variances of forecast data. More empirical data on actual project outcomes are necessary in order to establish a realistic picture of environmental impacts and to apply a stochastic risk approach to their assessment. Data from environmental auditing of the Great Belt, Øresund, Channel tunnel and similar projects will eventually begin to make such a stochastic approach possible. The approach should be kept simple, be based on actual experience and be transparent in order to support stakeholder and public participation in the decision-making process. In the following section we look at evidence from post-auditing at Great Belt and Øresund.

> The most pertinent issue regarding environmental impacts may be, not to predict impacts accurately *ex ante*, but to define appropriate environmental goals and then set up the organisation that can effectively adapt and audit the project to achieve the goals.

Towards better practice: Great Belt and Øresund

The Great Belt and Øresund links may, as already said, be considered among the most advanced major transport infrastructure projects in the world as regards the inexact science of environmental auditing and management. There is much to learn from these projects in terms of environmental protection for those involved in megaproject development, be they project promoters, governments, consultants, environmental groups or citizens. Experience from Great Belt and Øresund show that the most pertinent issue regarding environmental impacts may be, not to predict impacts accurately *ex ante*, but to define appropriate environmental goals

58   Megaprojects and Risk

and then set up the organisation that can effectively adapt and audit the project to achieve the goals in an ongoing process from project design through construction to operations. In what follows we focus on the Great Belt link because, at the time of writing this book, it was the only one of the two projects that had been completed long enough, since 1998, to allow full post-auditing. Data from Øresund indicate, however, that the results here resemble those from Great Belt, at least as regards marine environmental impacts.

When, in 1987, the Danish Parliament decided to build the Great Belt link, environmental groups had succeeded in placing certain restrictions on the project regarding environmental impacts. Most importantly, after the link was constructed the flow of water through the Belt had to be unchanged so as not to impact the fragile ecology of the Baltic Sea where even small changes in the salt balance were seen as possible causes of severe changes in the flora and fauna of the Sea. This would become known as the so-called Zero Solution. Furthermore, it was decided that biological impacts from the project must be monitored, controlled and minimised. Later the political parties behind the project decided on further restrictions on environmental impacts and the result was an incremental optimisation process in which the project design, with its two bridges and a tunnel, and the construction methods were adapted, step by step, to environmental demands. Some of the more important environmental adaptions were:[28]

- Choice of bored tunnel instead of immersed-tube tunnel;
- Replacing a 1,200 m ramp with a bridge. Shortening other ramps by 1,050 m;
- Streamlining of underwater foundations in order to reduce blockage and compensation dredging;
- Increasing the span of the suspension bridge in order to reduce the number of bridge piers and thus blockage;
- Fine-tuning the location, volume and timing of compensation dredging in order to minimise spillage and other negative impacts;
- Recycling of sediments for use in the construction of the link; depositing of several million tonnes of sediments in specially designed basins in an artificial extension of Sprogø island.

The total costs for adapting the project to environmental concerns and for the environmental monitoring programme have been estimated by the project company, Sund & Bælt, at approximately DKK 1.5 billion, or 7 per cent of construction costs (1988 prices).[29] The costs of the environmental monitoring programme were DKK 130 million (1988 prices), equivalent to DKK 210 million in 2001 prices. The figure for environmental costs does not include the costs involved in choosing a bored

Environmental impacts and risks 59

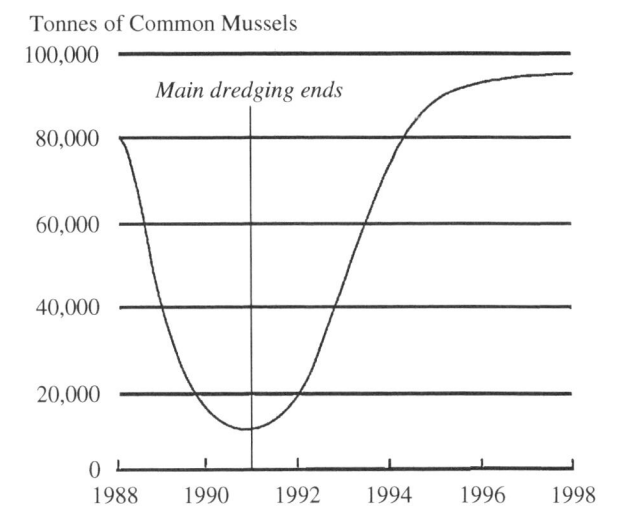

Figure 5.1 Impact of the Great Belt link on common mussels before, during and after construction
Source: Sund & Bælt, *Storebælt og miljøet* (Copenhagen: Sund & Bælt Holding, 1999), p. 51.

tunnel instead of an immersed-tube tunnel.[30] If the bored tunnel were included in environmental costs these would be substantially higher, not least because the tunnel turned out to pose more problems than expected, resulting in cost overruns for this part of the project of DKK 2.9 billion or 110 per cent (see Chapter 2).

In addition to the many technical–environmental innovations on Great Belt there is also reason to emphasise the organisational set-up of the environmental programme. Environmental concerns and environmental management were centrally and strategically placed in the overall project organisation and were adequately funded. In addition, an external panel of ten independent, international experts was used from 1988 to 1998 for review of biological and hydrographical issues. The independent panel helped reduce the risk of unexpected negative marine impacts occurring and also served to increase public trust in the marine environmental programme on Great Belt.

In 1999, after construction of the Great Belt link was completed, Sund & Bælt documented that the Zero Solution had been achieved and the documentation was accepted by the environmental authorities. The flow of water through Great Belt was considered unchanged.[31] As regards biological, marine impacts, the project company recently announced that flora and fauna, which were negatively affected during construction, have been re-established and that no substantial permanent impacts are expected from the construction of the link.[32]

On land, the most important environmental impacts from the Great Belt link have resulted from changed travel patterns. A shift from aeroplanes and ferries to trains and cars has occurred as a result of the opening of the fixed link. According to Sund & Bælt, this shift has produced yearly savings in energy consumption of 2.5 to 3 PJ (peta joules), or over 200,000 tonnes of carbon dioxide.[33] This is equivalent to two per cent of total energy consumption for transport in Denmark. A so-called life-cycle analysis shows that in terms of energy the payback period for the fixed link will be six to seven years, in other words after this time the energy used for building and operating the link will be more than offset by the energy saved. According to Sund & Bælt this leaves room for an increase in road traffic to approximately 30,000 vehicles per day before energy used is higher than energy saved.[34] This is three times the number of vehicles transported across Great Belt by ferry in 1998 before the link was opened to cars and 40 per cent higher than actual traffic in 2001. Therefore, according to Sund & Bælt, there is still room for increased road traffic across Great Belt before the positive impact on energy consumption from building the link disappears and turns into a negative environmental impact.

It is interesting to note, however, that a study published by the Danish Transport Council, an independent think tank set up by the Danish Parliament, reaches conclusions that are quite different from those of Sund & Bælt regarding impacts on energy consumption. According to this study, the payback period for the link in terms of energy will be 14 years instead of the 6 to 7 years found by Sund & Bælt. And a further increase in road and ferry traffic of only 15 per cent will make payback in terms of energy impossible, that is energy used will exceed energy saved.[35] The study concludes, 'the large reductions in energy consumption and carbon dioxide emissions from traffic that could have been achieved from the fixed link have not occurred. They have been eaten up by traffic growth.'[36]

A closer examination of the two studies reconfirms a general fact about megaprojects: rarely is there a simple truth about such projects. What is presented as reality by one set of experts is often a social construct that can be deconstructed and reconstructed by other experts. For Great Belt, the facts behind the facts show that the differences in results between the two studies of energy consumption are due mainly to different assumptions about car and ferry traffic, for instance regarding how many cars would travel across Great Belt in 1999 and how much energy the ferries would consume.[37] We conclude that transparency and the use of independent

Environmental impacts and risks 61

experts are as important for issues of energy consumption and global warming as they are for marine impacts. As such they should be an integral part of the organisational and institutional set-up, which they were not for the Great Belt link.

> Rarely is there a simple truth... What is presented as reality by one set of experts is often a social construct that can be deconstructed and reconstructed by other experts.

Despite such criticism, when summing up the experience from Great Belt as regards environment, the conclusion is that the project has been an important step in the right direction for megaproject development. Undoubtedly there is scope for further improvements, but the main lesson to be learned for other projects is that if environmental groups place sufficient pressure on decision makers at the outset, here the Danish Parliament, then it is possible to plan and implement major tunnels and bridges with due consideration for the marine environment, even in a sensitive habitat such as the Baltic Sea. During construction, negative environmental impacts were often larger than those forecast, but since construction was completed it appears that flora and fauna have been re-established quickly and that no permanent damage is expected. Preliminary data for Øresund indicate that the situation is similar here.

Experience from Great Belt also shows, however, that new transport infrastructure may generate so much new and unanticipated traffic that what was forecast to be a large positive environmental impact, here lower energy consumption resulting from reduced ferry and air traffic, risks becoming only a small improvement or even a negative impact. Traffic demand management based on environmental policy objectives is necessary if policy makers want to avoid this situation. Such management has not been applied for Great Belt; quite the opposite: the price for using the road link has been reduced successively and car traffic has instantly outgrown even the most optimistic forecasts (see Chapter 3).

At the time of writing this book, a high traffic volume was not a problem on the Øresund link. Here the most pressing issue was not environmental but financial: would traffic be sufficiently large to pay back the loans incurred to build the link? The situation for Øresund well illustrates the trade-off, or dilemma, between environmental and financial objectives for this type of megaproject: in order to finance the huge investments, the project company and its owners – here the Danish and Swedish governments – are dependent on a high volume of traffic, which runs counter to environmental policies aimed at reducing energy consumption and emissions of greenhouse gases. This is the dilemma faced by the Danish and Swedish governments today for the Great Belt and Øresund links. So far the governments have chosen in favour of increased traffic.

Finally, experience from Great Belt and Øresund shows that environmental protection comes at a cost and that this cost tends to translate into substantial financial risk. A pertinent but unanswered question in relation to cost is whether the hundreds of millions of kroners spent on environmental protection on the Great Belt and Øresund links would have bought more and better environmental protection if spent elsewhere on different measures, even if the objective were the specific one of protecting the marine environment of the Baltic Sea. This type of question remains unanswered not only for the Great Belt and Øresund links but also for other types of project and policy. Thus the burden of the question should not be carried by the Great Belt and Øresund links alone. The question must be answered, however, before we arrive at effective environmental policies.

> Experience ... shows that environmental protection comes at a cost and that this cost tends to translate into substantial financial risk.

The importance of post-auditing to environmental learning

Experience indicates that problems exist with the scoping, timing and reliability of environmental impact assessment of large transport infrastructure projects. The framework for environmental impact assessment used today needs to be extended and clarified, while the approach should be made more operational and should incorporate standard concepts of risk analysis and management. The following are needed:[38]

(1) To undertake the environmental investigations at an early stage and in parallel with other investigations;
(2) To establish mandatory scoping with specified procedures;
(3) To establish assessment criteria which can be expressed as performance specifications with respect to environmental impacts and risks (regarding the use of performance specifications, see also Chapter 10);
(4) To establish performance specifications that can be measured, monitored and audited by scientific methods and, if necessary, would be able to meet the rigours of a legal process;

(5) To allow for active participation by stakeholder groups and the general public in environmental impact assessment, monitoring and auditing, including the establishment of criteria and performance specifications.[39] (See Chapters 10 and 11 for more on participation of stakeholder groups and the general public.)

As part of such operationalisation, the following need to be taken into account:

(1) The details of the approach to environmental impact assessment should be fixed before any decisions are made regarding which technical solutions should be considered for further study;

(2) A uniform system of environmental impact assessment should be applied for all proposed technical solutions;

(3) Where relevant, a probabilistic approach should be used for assessment, including risk analysis of impacts, in order to increase reliability. However, such an approach should be kept simple, experiential and transparent in order to support stakeholder and public participation in the decision-making process;[40]

(4) Regional, systemic and long-term issues should be included, for instance by assessing the project in the context of other major projects, where relevant;[41]

(5) What has been called 'extended peer communities' should be developed and used in the assessment process. Hence independent expertise whose affiliations lie outside those of the experts officially involved in project development should be used for peer review of environmental impact assessment, monitoring and auditing.[42] Here it is possible to learn from the positive experiences of conducting such peer reviews in the Øresund and Great Belt projects;

(6) Requirements for detailed monitoring and auditing programmes should be included in environmental impact statements, including mechanisms designed for monitoring and auditing during possible project implementation and after project completion, as in the case of the Great Belt and Øresund projects;

(7) Monitoring and auditing should allow for transparency and participation by stakeholders and the public.[43]

In addition to these tasks, it is a key issue regarding environmental impact assessment, monitoring and auditing that an institutional framework should be developed that will effectively accomplish the above tasks. As pointed out by Wood, Dipper and Jones, the main obstacles to learning about actual environmental risks in planning projects are the absence of mandatory, institutionalised requirements for post-auditing and indifference among authorities, developers and the general public to such audits.[44] Such obstacles turn out to be particularly pronounced for one-off developments such as the typical megaproject, the Great Belt and Øresund links being notable exceptions. But with every development for which an environmental impact statement is submitted but no follow-up analysis is carried out – which is the typical situation – an important opportunity is being missed to learn about environmental risk and thus to mitigate one of the most important types of risk in risk society.

> The main obstacles to learning about actual environmental risks are the absence of mandatory, institutionalised requirements for post-auditing and indifference among authorities, developers and the general public to such audits.

6 Regional and economic growth effects

In this chapter we focus on regional and economic growth effects of megaprojects. As in the previous chapter, our main purpose is to identify past problems that may prove useful in understanding and improving the decision-making process for such projects.

Recent years have seen a resurgence of interest in the impact of infrastructure on regional development and economic growth. Indeed, one of the arguments often advanced for committing public funds to infrastructure investments is that it will generate economic growth in general, in a region or a country, and/or in a particular local area.[1] There are good theoretical and empirical reasons for approaching such claims with caution, we will argue. A hard-nosed approach to this subject is often warranted since much of the interest in various circles for infrastructure investments reflects rent-seeking behaviour. Such behaviour is explained by the circumstance that infrastructure investments may generate benefits to specific construction and user groups while the major part of costs is often borne by the taxpayers.[2]

> One of the arguments often advanced for committing public funds to infrastructure investments is that it will generate economic growth. There are good theoretical and empirical reasons for approaching such claims with caution.

Transport infrastructure and economic development

Both business and private persons use transport infrastructure. In general, infrastructure must be seen as an input into the production of a transport service, which in turn is used as an input into a final product or service demanded by consumers, such as a visit to a relative or the availability of a certain commodity in the local store. The economic effect of investments in transport facilities is a decrease in the cost of the input, either through decreased outlays for producing a trip or decreased time spent on a trip, which may also translate into reduced outlays, or both. In short, the investment reduces the 'price', or the generalised cost (including expenditures, time, inconvenience and other factors), to users availing themselves of those services that the infrastructure produces.

From a theoretical point of view, the lowering of the price of infrastructure services will have two effects in the short term. First, it will increase the demand for that service by both private individuals and businesses, and secondly, it will increase the level of profits made by the businesses. For these, the effect will be rather limited in the short-term as the profit increase primarily comes from a substitution of transport via the new infrastructure for other factors of production (the effect from a change in the relative prices of factors of production). This short-term effect on business will be reflected in the national accounts as a contribution to economic growth.

In the medium term there will be an additional effect. Businesses that were already making use of the infrastructure before its upgrading (or an alternative substitute infrastructure) can be expected to increase their demand further as the prices of the goods or services they produce can be reduced, thus leading to a change in spatial economic interactions. However, a precondition for this to happen is that the goods and services are subjected to competition. While the shorter-term effects are, in principle, fairly easy to determine, longer-term effects are more difficult to predict. The reason for this, as concerns households, is that these effects are determined by the limited availability of time and land. If, for example, a new bridge can provide access to new areas where land is plentiful and therefore cheap, without adding much additional time for commuting or travelling for other purposes, the lowering of the cost of making use of the infrastructure may give rise to relocation effects. In the vicinity of larger conurbations, such effects may be considerable, particularly from the point of view of the number of trips recorded on the new infrastructure.

For businesses, the long-term effects are determined by the competitive forces that are generated by the fact that the profit level is increased initially within those companies that immediately benefit from the new infrastructure. What actually happens depends on the relative scarcity of land, labour and capital. Assuming that land is plentiful, then expansion, through relocation or through further additions to capacity, will take place. There will be further economic growth in certain regions, and maybe also in a country as a whole. If it is assumed, on the other hand, that land is not plentiful, then the only consequence of the infrastructure is that land values will increase. However, while only the former of these two alternatives will result in new investments being made and in increased employment, both of the alternatives will be recorded as contributing to economic growth as measured by the national accounts, and to economic growth both for the country as a whole and for the regions that benefit from the investment in the infrastructure. Hence, only under specific circumstances will that growth be accompanied by increased employment and investment in capital and facilities. Moreover, a precondition for overall economic growth is that the infrastructure in question is economically viable and generates an economic rate of return above a certain level.

The question is, then, whether a traditional cost–benefit analysis will be able to catch all these effects leading to economic growth, and possible relocation and expansion effects in certain areas. The answer to this question is, in principle, yes, even if it may be difficult or impossible in practice. If an economic (or social) cost–benefit analysis has been carried out in an appropriate way, in principle the result of the analysis will reflect all the real effects of the investment at the local level.[3] This is a reflection of the condition that the economic benefits of an infrastructure investment can always be measured in two equivalent ways, namely either by looking at the change in the 'income statements' of all people and all businesses or, in the case of transport infrastructure, by looking at the effects on the traffic that can be expected to make use of the infrastructure, in other words what is 'on the infrastructure'. In the latter approach, which is easier to implement, the regional relocation effects have to be integrated into the traffic forecasting procedure.

In the case of market imperfections (for example: monopolistic competition) the overall regional economic effects might deviate from those measured by conventional cost–benefit analysis (additional surpluses of old and new users after investment). In a paper prepared for SACTRA (Standing Advisory Committee on Trunk Road Assessment) of the UK Department of Transport, Venables and Gasiourek work out several cases for such additional benefits to occur in an imperfect market environment.[4] The equilibrium analysis is purely theoretical and has not been empirically tested. But it nevertheless gives reason to study the impacts of large infrastructure investments and of investment programmes on sectors that are not directly affected by the reduction of transport costs but benefit indirectly from lower prices induced by higher competition (forward/backward linkage effects). It should be noted in this context that the theoretical study of Venables and Gasiourek clearly leads to the conclusion that the additional effects that are not measured by conventional cost–benefit analysis might be positive for some regions and negative for others. Therefore, to measure the magnitude of such effects is not the only issue; it is even more important to measure the direction and the induced regional and sectoral feedback mechanisms.

> Even for a giant investment such as the Channel tunnel, five years after the tunnel's opening there were very few and very small impacts on the wider economy. The potential impacts on the directly affected regions were found to be mainly negative.

Empirical evidence

In view of the fact that the Channel tunnel and the Great Belt and Øresund links only recently came into operation, no full *ex post* studies of the development effects of these links are available. A preliminary study exists of impacts from the Channel tunnel, especially in Kent, five years after the tunnel's opening. In addition, a number of *ex ante* studies have been carried out. The studies indicate the following.

An early *ex ante* study of the Channel tunnel, undertaken by Roger Vickerman, reviewed the impact primarily from the point of investment and employment generation in various regions in England. The study is based on an analysis of:
 (i) the potential cost savings created by the Channel tunnel;
 (ii) the importance of transport in total industry costs; and
 (iii) the potential for realising cost savings.

The conclusion of the study is, there 'are no compelling reasons for believing that the Channel tunnel project will create an economic bonanza for the adjoining regions. If anything, there is some evidence that benefits are more likely to accrue to locations at some distance from the Tunnel itself, say 100 to 150 km.'[5] The study further concludes, 'there is little immediate gain to be made from the Channel Tunnel project either for the local economies at the Tunnel-head or for the national economies'.[6] The study concludes that the possible regional development effects, which could be generated, would only come about as a consequence of appropriate further complementary investments (see further below).[7]

The first *ex post* study of actual impacts from the Channel tunnel, also by Vickerman, confirms the earlier observations. The study concludes that even for a giant investment such as the Channel tunnel, which is several times the size of most megaprojects, five years after the tunnel's opening:
 (i) 'there are very few and very small impacts on the wider economy'; and

(ii) '[i]t is difficult to identify major developments associated with the tunnel'.[8]

The potential impacts on the directly affected regions are found to be 'mainly negative'.

The Great Belt link has been subjected to a considerable number of studies with various techniques employed. In a report from the Danish Transport Council, the predictions of these studies are summarised as follows:[9]

(1) The effects on the port cities that will lose ferry operations will be negative;
(2) The regional economic effects are assessed as small;
(3) The positive economic benefits that could be generated would accrue to a sizeable area from Copenhagen in the east to southwestern Jutland.

After more than two years of operations, it appears that no major negative effects from the fixed link are evident, apart from the direct effect of closing down the ferries. No major positive effects appear to be manifest either, but no actual *ex post* study has been carried out. The *ex ante* studies summarised by the Danish Transport Council expect the positive economic benefits mainly to accrue to existing urban areas. Similar results have been derived in other studies concerning infrastructure projects.

For the Øresund link, two *ex ante* studies are available, both of which are of a qualitative nature. These studies place emphasis on the lack of integration between the Swedish and Danish economies in the vicinity of the fixed link, and therefore identify a considerable scope for further integration of the economies of the two countries flowing from the establishment of a fixed connection.[10] One of the studies concludes that the link could result in an increase in the value of production in the agglomeration established through the link by connecting the two sides of Øresund in the amount of DKK 20 billion.[11] However, no detailed analysis of how this would come about has been presented, and it is also unclear whether this increase is explained by the fixed connection alone, or would partly come about even without this.

Although not based on in-depth regional impact studies, the city administration of Malmö is convinced that the fixed link will have a positive effect on regional development. Action has been taken with respect to land-use development and linking the city transport network to the access links to the Øresund bridge. Expectations are that people might move from the densely populated Copenhagen region to Malmö, followed by new market businesses. A university in Malmö will be integrated in an Øresund education network. On the Copenhagen side there is an effort to develop a new town on virgin land, the so-called Ørestad, complete with offices, commercial space and homes, and with high accessibility from both the Øresund link and Copenhagen's city centre. The expectation is that development will be stimulated by future integration effects. These activities on both sides of the Øresund link are hoped to create bootstrap effects or some type of self-fulfilling prophecy, which are then intended to multiply the initial effects of the project.

> The Øresund link is currently best seen as a grand social experiment in cross-national and cross-cultural integration via transport infrastructure development. Such experiments are rare.

Available studies indicate that major transport infrastructure investments, which connect nearby major urban areas and thereby cause a significant reduction in travel time, can have a substantial impact on economic activity. This observation is supported by the developments subsequent to the construction of the bridges across the Bosporus to link Istanbul with Asia Minor and across the Faro river in Portugal.[12] At the same time caution has to be advanced since the Øresund link, unlike the other links, connects two countries, and is, furthermore, not located on the route that provides the shortest distance between Sweden and continental Europe from the point of view of interregional or long-distance traffic. It is a longstanding experience documented by geographers around the world that inducing integration and development across national boundaries is a reticent and slow process. This is the case also where two nations speak the same language and there are no separating waters or other natural barriers as with, for instance, Canada and the USA. Studies of spatial interaction have shown that even for a relatively friction-free activity such as telephone calls, the border between Canada and the USA forms a major barrier.

In terms of spatial interaction, nations, quite simply, tend to turn their backs on each other, that is interactions within a nation are typically substantially more intensive than interactions between nations. A recent study of cross-border networking activities between two neighbouring regions in Germany and France confirmed this observation once again and concluded that 'spatial distance does not matter, cultural and institutional distance does'.[13]

After connecting Denmark and Sweden in 2000, the Øresund link has delivered additional evidence in support of this conclusion, and perhaps it is still too early in the life of the link to expect anything else. Institutional barriers originating in different tax, employment and welfare systems in the two countries have been underestimated and have impeded regional and economic integration. As a result, and as mentioned in Chapter 3, traffic on the link has been lower than forecast. Traffic has also been unbalanced in the sense that it has been dominated by Swedes visiting Denmark. With a reference to Sweden's strict liquor laws and the free availability of alcohol in Denmark, this has given rise to an unfair but popular joke describing the Øresund link as the most expensive trail to a watering hole ever built. Joking aside, the Øresund link is currently best seen as a grand social experiment in cross-national and cross-cultural integration via transport infrastructure development. Such experiments are rare and only time will tell whether this one succeeds or not. Thus development in the Øresund region deserves close scrutiny in the years to come.

In sum, studies of the effects of investments in transport infrastructure on regional development and growth generally come to the conclusion that the argument for transport investment achieving regional economic growth effects is weak and that 'those analyses that have surveyed the indicators of regional development are unanimous in their findings that the effects of even relatively large infrastructure projects are marginal'.[14]

These studies also identify the main reason why this is so, namely that the transport costs are a relatively small component of the final price of goods and services, say 1 to 7 per cent.[15] Even large-scale investments that give rise to large time savings thus have a very small impact on the income statement of companies. A much-quoted study regarding the effects of the M62 motorway, from Liverpool via Manchester to Leeds and Hull, in England found that the reduction in transport costs meant only a 0.33 per cent decrease in the production costs of affected firms, and that only 2,900 jobs per year would be affected as a result of the new facility out of a total labour force in the area of 3.4 million.[16]

The available knowledge suggests that only under a number of specific conditions can one expect a significant positive impact of investments in transport infrastructure on regional economic growth. These are:[17]

(1) If serious capacity problems exist in a network in a region, investments to remove the bottlenecks are likely to affect investments and employment in the region;
(2) Investments in large urban areas, where the new capacity affords significant savings in relation to the cost of transport, are likely to give rise to relocation of both households (in the shorter-term perspective) and of companies (in a longer-term perspective);
(3) When there is a combination of various types of investment in both infrastructure and social capital.

Condition (3) has been observed, for example, in the case of airport provision which turned out economically successful when the transport investment was accompanied by proactive regional development policy to attract new business and leisure activities.[18]

Lessons of growth effects

The main lessons to be learnt regarding regional and economic growth effects are three:

(1) It is common for proponents of major transport infrastructure projects to claim that such projects will result in substantial regional and/or national development effects. Empirical evidence shows that such claims are not well founded, the main reason being that in modern economies, transport costs constitute a marginal part of the final price of most goods and services;

(2) Investments in transport infrastructure may be expected to result in significant development effects mainly in the following situations:
    (i) in regions where serious capacity problems exist in a network;
    (ii) in large urban areas where new capacity would result in significant savings in the cost of transport; and
    (iii) in situations where investments in various types of infrastructure go together with investments in social capital;

(3) In terms of risk, regional and economic growth effects are not the most pressing concern for major transport infrastructure development, unless such development has been oversold by promoters with the notion that it would generate substantial growth. Overselling of that kind may in itself be a risk. Highly localised negative employment effects may also pose substantial risk to specific communities and may need to be compensated.

7 Dealing with risk

The importance of risk analysis

As mentioned in Chapter 1, in terms of risk, too many feasibility studies and appraisals of megaprojects assume projects to exist in a predictable Newtonian world of cause and effect where things go according to plan. In reality, the world of megaproject planning and implementation is a highly stochastic one where things happen only with a certain probability and rarely turn out as originally intended. The failure to reflect the probabilistic nature of project planning, implementation and operation is a central cause of the poor track record for megaproject performance documented above.[1]

It is fairly common in feasibility studies and appraisals of major transport and other infrastructure projects to make a mechanical sensitivity analysis examining the effect on project viability of hypothetical changes in, for instance, construction costs, interest rates and revenues.[2] The typical range for such sensitivity analysis is from ±10 per cent to ±20 per cent. It is, on the other hand, unfortunately rare that risk analysis is made by identifying alternative future states of costs, revenues and effects and a probability distribution estimated for the likelihood that these states would actually occur. This information is required in order to estimate the expected values of costs, revenues and effects, or, in other words, the most likely development, including the associated variances. Approaching risk analysis in this way is essential in order to curb what has been called 'appraisal optimism' and to give decision makers a more realistic view of the likely outcome of projects, instead of the incomplete and misleading view on which decisions are often based today.[3] Risk analysis is also the basis for risk management, that is the identification of strategies to reduce risks, including how to allocate them to the parties involved and which risks to transfer to professional risk management institutions, namely insurance companies. In this chapter we describe how risk was treated in the Channel, Great Belt, Øresund and other projects. In addition, we review the implications of risk for costs and financing. Finally, we spell out the lessons to be learnt regarding risk.[4]

Channel tunnel, Great Belt, Øresund and other projects

When Eurotunnel went public as a company in 1987, investors were told that the project was relatively straightforward. Regarding risk, the prospectus read:

> Whilst the undertaking of a tunnelling project of this nature necessarily involves certain construction risks, the techniques to be used are well proven... The Directors, having consulted the Mâitre d'Oeuvre, believe that 10%... would be a reasonable allowance for the possible impact of unforeseen circumstances on construction costs.[5]

Two hundred banks communicated the figures for cost and risk to investors, including a large number of small investors. As has been observed elsewhere, anyone persuaded in this way to buy shares in Eurotunnel in the belief that the cost estimate was the mean of possible outcomes was, in effect, misled.[6] The cost estimate of the prospectus turned out to be a best possible outcome based on the unlikely assumption that everything would go according to plan with no delays, no changes in performance specifications, no management problems, no problems with contractual arrangements or new technologies or geology, no major conflicts, no political promises not kept, and so on. The assumptions were, in other words, those of an ideal world. The real risks for the Chunnel venture were several times higher than those communicated to potential investors, as evidenced by the fact that the real costs of the project were higher by a factor of two compared with forecasts.

Similarly, before ratifying the Great Belt project, the members of the Danish Parliament were informed by the Minister of Transport regarding risk that:

> I do not consider the uncertainty of the overall construction cost estimate for a fixed link across Great Belt to be larger than for other large bridge or tunnel projects carried out in this country.[7]

However, the cost of even the largest bridge and tunnel projects carried out in Denmark before the Great Belt link was less than a tenth of the original Great Belt budget, and none of these projects included a bored tunnel.[8] So for reasons of size and innovation alone, the risks associated with the Great Belt link were more important than for any other transport infrastructure project in modern Danish history. In addition, the geological and technological risks – and hence the risks associated with construction costs – were higher. Nevertheless, no financial or economic risk analysis was made for the Great Belt project.[9] As in the case of the Channel tunnel, the cost estimate for the Great Belt link turned out to be optimistic and was closer to an unlikely best possible outcome than to a most likely one. With a cost overrun of 110 per cent for the Great Belt rail tunnel, risks caught up with the rail link, which proved non-viable and was rescued only by cross-subsidisation from the road link (see Chapter 4). Total construction costs were 54 per cent higher than forecast costs. But, unlike the Chunnel, it is taxpayers' money that has been placed at risk at Great Belt and not investors' money.

For the Øresund link, partial risk analyses were carried out. In one such analysis, a group of government officials assessed that given historical experience a 50 per cent cost overrun 'cannot be seen as an unrealistic maximum estimate' for the link.[10] In addition, the group found the project financially non-viable even without this overrun, as did three other appraisals carried out by officials in the months before the project was presented to the Danish Parliament. Yet, when the Minister of Transport presented the project for ratification in 1991, none of this information was mentioned. Neither the proposed law nor the accompanying comments contained any information about risks of non-viability.

When, more than two years later, it became publicly known that such information existed and had been withheld from Parliament, the result was a sharp critique of the Minister of Transport by the Auditor-General of Denmark.[11] The Auditor-General found that the assumptions on which the estimates of project viability were based 'represent the economic assumptions that have to be made for the project to be self-financing', that is the assumptions that would make the project seem viable on paper, and not the assumptions that would have reflected the most likely development.[12] For later budgets, worked out by the state-owned enterprises responsible for construction of the Danish access links and the fixed link proper, the Auditor-General similarly found that it was 'less evident' that the budgets were within claimed levels of uncertainty.[13] Moreover, the Auditor-General found that the risks regarding the future development of interest rates and financing costs had been ignored and that it 'would have been natural' to mention this in the project budgets.[14]

On this basis, the Auditor-General resolved to monitor the Øresund project and to carry out audits in order to establish whether the assumed basis for the project would actually be realised, including whether the project would, in fact, be self-financing as required by the Danish political agreement behind the project and by the agreement between Denmark and Sweden to build the link.[15]

The treatment of risk in the Channel tunnel, Great Belt and Øresund projects has clearly been inadequate. Even so, the organisational and institutional set-up of the Great Belt and Øresund links, as loan- and user-financed state-owned enterprises, may be improvements over earlier transport infrastructure projects in Denmark in terms of risk identification and allocation, because costs, revenues and viability have become more visible than in earlier projects, as have environmental impacts. Still, there is substantial scope for improvement for future projects as shown above and as will become clear from what follows. For other major projects – transport and non-transport alike – the conclusions are similar: the risks involved are high and are typically treated in a deficient, sometimes even deceptive, manner in feasibility studies and project appraisals, if treated at all. In a World Bank study of ninety-two projects, only a handful was found to contain 'thoughtful' risk analysis showing 'good practice'.[16] Appraisals of World Bank projects are typically more complete and more rigorous than appraisals of other projects.

Nevertheless, it is important to note that there are good examples as well. We have already mentioned in Chapters 2–4 the construction of the Paris South-East and Paris Atlantique high-speed rail lines.[17] Also, the technologically high-risk Apollo aerospace programme is considered a classic success story of megaproject planning and implementation. The cost overrun on this US$21 billion project was only 5 per cent. Few know, however, that the original budget estimate included US$8 billion of contingencies.[18] By allowing for risk with foresight, the programme avoided ending up in the type of large cost overrun that destabilises many major projects during implementation. The Apollo approach, with its realistic view of risks, costs and contingencies, should be adopted in more major projects.

> In a World Bank study of ninety-two projects, only a handful was found to contain 'thoughtful' risk analysis showing 'good practice'.

A typology of risks

The main sources of financial risk in major transport infrastructure projects are:
(1) construction cost overruns induced by, for instance, government, client, management, contractor or accident;
(2) increased financing costs, caused by changes in interest and exchange rates and by delays; and
(3) lower than expected revenues, produced by changes in traffic volumes and in payments per unit of traffic.

Although less significant, financial risks are also related to costs of operations, maintenance and management. From an economic point of view the main risks are cost overruns, delays and lower realised demand than that assumed during appraisal.

From an analytical point of view, it is expedient to identify the following types of risk of relevance to both a financial and an economic perspective:[19]
- Project-specific risks;
- Market risks;
- Sector-policy (including *force majeure*) risks;
- Capital-market risks.

The two first types of risk include those that are conventionally associated with a project, and that have been in focus so far. As for project-specific risks, the conventional assumption is that their effect can be eliminated – at least in part – by risk pooling or risk spreading; see further below. This does not normally apply to market risks that are explained by more fundamental events that affect economic activities in a similar way, for example the overall economic development in a country. Sector-policy risks arise from the fact that the outcome of a project is dependent on specific sector policies; for transport projects, for instance, complementary investments in access links, taxation of transport or other regulation of road transport or of the environment. Some of these risks can be identified and can also be eliminated by providing a stable regulatory environment and by proper contracting. These types of risk are not necessarily eliminated by having projects operated by state-owned companies. Ultimately, such companies serve the general public, and if changes in regulatory policies imply that a project cannot be used as originally envisaged, private parties, that is taxpayers, will be affected in a negative way, as would be the case if the project were to be operated by a private entity.

Capital-market risks are created through borrowing, in particular in the international market, in order to finance projects. Such risks mainly consist of two elements, interest rate risks and currency risks. The capital market is today able to provide financing on conditions that allow borrowers to protect themselves against currency and interest rate adjustments in the short and medium term. But, of course, such insurance comes at a cost.

The cost of risk

The condition that a project is associated with risks gives rise to an economic cost. People are normally risk-adverse and are prepared to pay something – an insurance – in order to reduce or totally eliminate risks. The cost of risk is an economic concept and reflects the maximum amount that an individual is willing to pay to eliminate a particular type of risk, so that the future of a particular type of event would become risk free for that individual.[20] In practice, different levels of riskiness associated with different types of investment are reflected in different minimum rates of returns, which are required in order to persuade individuals to commit their money.[21] The lowest return is normally required on government bonds; here the return is about 3–4 per cent in real terms, as these bonds are considered to be virtually risk free. Private-sector debentures are normally associated with a somewhat higher return, say 5–6 per cent in real terms. The return on equity, namely ownership capital with risk, starts at about this level, but may become much higher depending on the perceived riskiness of the enterprise concerned.[22]

The risk costs associated with investments in infrastructure can be expected to be high. There are two basic reasons for this. The first is the fact that an investment in a major infrastructure project is basically a sunk cost, that is it cannot be retrieved. Once, for example, a bridge has been built it cannot really be used for anything else, so if the decision to build the bridge turns out later to have been a poor one, it cannot be repaired. The second reason is that the benefits of investments are often highly correlated with economic growth. If economic growth is high, then the project will fare well; conversely if growth is poor, the project will perform poorly. As noted, overall economic performance affects the market risk of a project, particularly when seen in an economic perspective. From an economic viewpoint it is difficult to offer insurance against market risks within a country, as it affects everybody, that is the risks cannot be spread and pooled when seen in a national perspective.

That infrastructure investments are viewed as risky is also brought out by the experience of private-sector involvement in infrastructure projects through a concession. The available data suggest that financing for concession companies will only be forthcoming provided: (i) the equity makes up about 20 per cent to 30 per cent of the total financing requirements; and (ii) that those who invest in equity can be expected to receive a return of between 15 and 25 per cent in real terms and after tax.[23] Assuming that the remainder of the capital, the long-term debt, can be mobilised at a real cost of 6 per cent, the implication is that the project will have to achieve a financial rate of return of about 7.5 per cent to 12 per cent in real terms (disregarding taxation effects).

The difference between the return required for this type of investment, say 9 per cent, and the return on a virtually risk-free investment, say government bonds at 4–5 per cent, can be viewed as a measure of the cost of risk associated with the project. It should be emphasised that this

Dealing with risk

type of cost is incurred not only when an infrastructure project is developed through a private concession company. It is also present when the project is developed by a state-owned enterprise and the financing is secured through a sovereign guarantee. In the world of risk there is no such thing as a free lunch. The risk and its costs exist under any circumstances, even if promoters of projects backed by sovereign guarantees often impart the impression to politicians and the general public that this is not the case, as happened for the Øresund and Great Belt links. Here the promoters seemed to hold this belief themselves. One high-ranking Danish Ministry of Transport official expressed it to us in the following way in his comments on an earlier draft of the manuscript for this book:

In my view you should mention [in the book] that the costs the private sector would demand to have covered for taking on the full risk [of projects] would make the projects more costly. Thus the advantage of financing the Great Belt and Øresund projects by a sovereign guarantee is that this lowered financing costs.[24]

Not only do we hold that sovereign guarantees hide financing costs instead of lowering them. We also hold that there are reasons to believe that the risk costs associated with financing could increase if the project is underwritten by a sovereign guarantee. The reason is that the guarantee will transfer most of the risks to taxpayers, that is risk bearers who can be assumed to be less able, on average, to protect themselves against risks than those persons who act in the capital market, hence increasing overall risk costs.

In addition, there is good reason to mention a point made by the World Bank that the money saved by lower interest rates on loans backed by sovereign guarantees may be offset by inefficiencies arising from relaxed discipline as a result of such guarantees. Lenders backed by sovereign guarantees have no, or much less, incentive for supervising projects than do commercial banks without such guarantees, resulting in relaxed pressure on project performance. According to the Bank it may take several percentage points of interest advantage to offset even moderate inefficiencies in terms of cost overruns and delays stemming from the inefficient supervision of projects.[25] For the Great Belt link the direct interest advantage of a sovereign guarantee on loans is approximately 1 to 2 percentage points, depending on the size and riskiness of the non-guaranteed projects that one compares it with.[26] For the Øresund link the interest advantage is of a similar size. If one were to compare interest on guaranteed loans against return on equity, the advantage would be higher.

The total risk costs associated with infrastructure projects can normally be expected to be substantial. Take, for instance, the Øresund link, which cost some DKK 17 billion (excluding access routes), and is expected to yield a financial return of about 5 per cent. Assume, for example, that the actual requirement is an additional 5 per cent, that is in total 10 per cent in order to ensure that the costs of risk also will be compensated for. Then the cost of risk is equal to about one third of the total investment, that is about DKK 6 billion. Whilst this example is not an exact calculation of the cost of risk, it illustrates how important it is to take risk into account in project planning and appraisal.

Strategies of risk assessment

The most consequential problem regarding risk analysis in megaproject feasibility studies and decision making is not the absence or inadequacy of risk analysis in itself, but the neglect of relevant downside probabilities in the calculation of project viability.[27] To a wide extent, risk is simply disregarded in feasibility studies and appraisals by assuming what the World Bank calls the EGAP-principle, 'Everything Goes According to Plan'.[28] We explained above how the unrealistic EGAP-principle was used for the Channel, Great Belt and Øresund links and what the problems involved were.

A cure recommended here, when undertaking a feasibility study, is to substitute what we call the 'MLD-principle' for the EGAP-principle, MLD standing for 'Most Likely Development'. The cure should be seen as part of a wider strategy for public-sector involvement in developing megaprojects, which is taken further in Chapters 10 and 11. By following the MLD-principle, the roles of feasibility study and appraisal are redefined from the optimistic and unrealistic everything-goes-according-to-plan estimation of project viability to the realistic and experience-based assessment of the most likely development of projects. Carrying out MLD appraisals, the focus is on identifying the most likely risks and the most risky parts in a given project in order to reduce these risks and, if possible, drop those parts. What the World Bank calls 'switching values' would be calculated for key variables, including environmental variables, understood as the level of the variable at which the project turns from viable to non-viable. The likelihood of switching values actually occurring would be estimated. In addition to this, what the Auditor-General of Sweden calls 'threshold levels' would be established for costs, revenues, environmental impacts and viability, namely levels that, if crossed, redefine the project as a new project that must be appraised and approved anew. The result of such measures would be more robust projects.

A further technique that should be made use of in feasibility studies sponsored by public-sector organisations is the analysis of worst-case scenarios, a method frequently employed in the private sector. The basic idea is simple: identify negative conditions from the point of view of the project and analyse the implications for the project's viability and financing. This approach is helpful for determining the robustness of the project, but also for identifying supplementary actions required in order to mitigate risks and ensure success. Worst-case scenarios may also be useful for identifying projects that should be dropped altogether since the risks and their implications appear to be all too significant.

Risk management

In addition to identifying risks carefully and making them visible, a main instrument for reducing the costs of risk is to prepare a risk management plan as part of a feasibility study. The purpose of such a plan is to identify how various risks are to be managed and by whom. In the public sector, the establishment of a credible risk management plan should be a part of the documentation required before any decision is taken on whether to go ahead with a project or not (see Chapter 11).[29] Figure 7.1 shows the main elements involved in risk management.

The main challenge to the preparation of a risk management plan is to actually fully identify the scope for risk management, and to communicate that it is much wider than what is normally appreciated. To a large extent this lack of appreciation of the role and scope of proper risk analysis and management is due to the history of contracting for infrastructure facilities. Because contracting in the infrastructure sector has to a large extent been on behalf of the public sector, the contracting format has become dominated by public-sector thinking. A key aspect of public-sector behaviour is thus that it is typically rule-based and not performance-based. In contracting, this is reflected in the fact that contracts typically are based on technical specifications being given to the contractor; the contractor's task is to build according to these specifications and not necessarily to achieve a certain level of performance (for so-called build contracts, see Chapter 9). However, this also means that the incentive and scope for developing new techniques in order to reduce costs, to reduce certain types of risk, and so on are limited. Consequently, present contracting techniques to a considerable extent eliminate the possibility of managing risks. A main challenge to risk management will therefore be to change the present contracting format, as is further discussed in Chapter 10.

There are several basic approaches to be considered as part of a risk management plan, some of which are partially overlapping. One approach

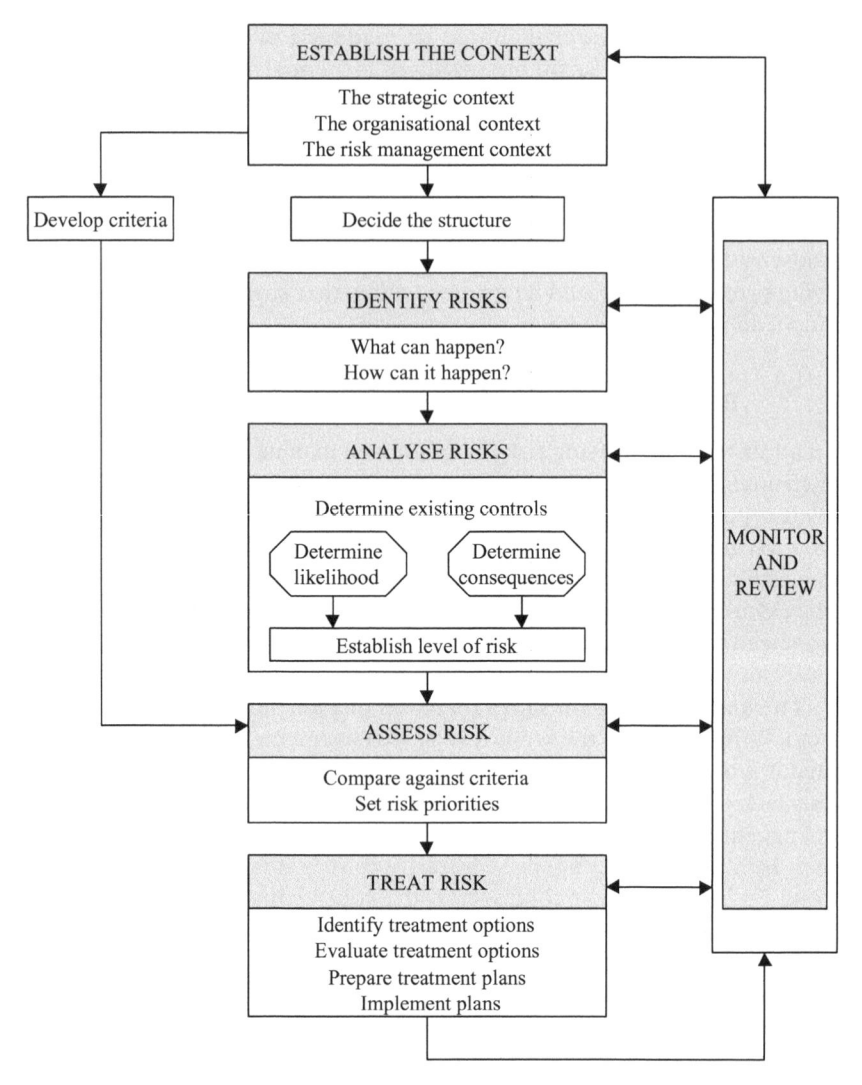

Figure 7.1 The risk management process

Source: Council of Standards Australia and Council of Standards New Zealand, *Risk Management*, AS/NZS 4360:1995 (Homebush: Standards Australia and Wellington: Standards New Zealand, 1995), p. 11.

involves eliminating risk altogether. This applies, for example, to certain sector-policy risks, for which it, given the circumstances, may be possible for the entity responsible for undertaking a specific type of infrastructure project to enter into an agreement with the central or local government to ensure that certain policy actions are not taken, or at least not taken without compensation. A second approach involves buying risk management services. This may be the approach used in order to deal with capital-market risks and it might also be available for what some consider as *force-majeure* risks.

A third approach involves allocating risks to parties who have an incentive to reduce the negative impacts of risk, either by reducing the likelihood of the event or to reduce the negative impact itself, if the event were to occur. Again this applies to such events that are normally considered to be of a *force-majeure* nature, although they may not necessarily be a major event. An example may be the occurrence of unexpected geological phenomena creating delays in the construction of a bridge or a tunnel. By allocating geological risks to contractors from the outset the result is likely to be a more thorough analysis of such risks before contracts are finalised and a faster and more effective containment of negative impacts should unexpected geological phenomena occur during construction. If, conversely, the owner accepts geological risks, or if the placement of such risks is unclear, experience shows this to be a sure road to delays and cost overrun, especially for projects with substantial underground work. Another example may be the occurrence of protest actions during the initial construction phase of an infrastructure facility, where the likelihood of the actions actually taking place can be significantly reduced through specific measures, for instance by replacing the conventional, closed format for megaproject development with a more open and transparent one and by paying adequate compensation to parties negatively affected by projects, as done in Boston's Big Dig project.

The nature of project-specific risks is such that their costs can be eliminated by appropriate pooling or risk spreading. There are several institutional arrangements to handle this. As has been demonstrated by Arrow and Lind, one approach to achieving effective elimination through risk spreading is to allocate project-specific risks to the public in general.[30] A way to achieve this would be to operate the project as part of the public sector, or by securing the financing for the project by way of government guarantees. But also the private sector has instruments available to achieve risk spreading. One of the reasons for establishing specific project companies for undertaking large infrastructure projects is thus to enable widespread participation by the capital market in the project, thereby allowing individual investors to pool their investments, and allowing the specific risks of the project considered to be spread between many investors, and thereby permitting the aggregate cost of this risk to be reduced.

The most difficult risks to manage are, as mentioned, market risks. One reason is that such risks are quite different when seen in an economic perspective from when seen from a financial point of view. As a rule, the economic cost of a market risk cannot be managed; the main issue to be considered is who should bear the cost of this risk, which is an important income distributional question and may also have institutional implications (see further Chapter 9).

> Public and private investors, parliaments, media and the general public are routinely inadequately informed and misled regarding the risks involved in megaprojects.

Lessons regarding risk

As documented in this and the previous chapters, the risks associated with major infrastructure projects are substantial. Key factors contributing to risk are the facts that the investment will be irreversible and the viability highly dependent on general economic development. Given the magnitude of the uncertainties involved, feasibility studies of major projects without risk analysis are less than useful since such studies will often deceive decision makers and the general public regarding the outcomes of projects. Risks cannot be eliminated from major projects, but they can be acknowledged and their impacts reduced through careful identification and by allocation of risks to those best suited to manage them.[31]

In most democracies the civil service has an obligation, defined by law, to provide the Cabinet and Parliament with 'all relevant information' pertaining to their decision taking and law making. Clearly, risks of cost overruns of 50–100 per cent on multibillion-dollar projects, together with large uncertainties regarding revenues and environmental impacts, are 'relevant information'. Thus such information must be brought to the attention of politicians and the general public.

The following conclusions can be drawn regarding risk:
(1) Public and private investors, parliaments, media and the general public are routinely inadequately informed and misled regarding the risks involved in megaprojects, cases in point being the Channel tunnel, Great Belt and Øresund projects;
(2) A full risk analysis based on the MLD-principle (Most Likely Development) should be carried out as part of feasibility study and appraisal – undertaken by public-sector organisations – for any megaproject. In addition, a risk management plan should be prepared. Such risk

analysis and management would identify the most risky parts of a project. The objective is to reduce risks and to change or drop the most risky parts of the project. Finally, the aim is also to allocate risks appropriately to the involved parties;

(3) Risk analysis should also comprise worst-case scenarios, in order to illustrate what happens if worst comes to worst. The experience with flooding and fire in the Great Belt rail tunnel illustrates the pertinence of this point as do the cost overruns and fire in the case of the Channel tunnel;

(4) Feasibility studies and risk analyses for future projects should be carried out together with considerations regarding the possible institutional, organisational and financial set-ups for the project. The set-ups will substantially influence risks and costs, just as risks and costs may influence the set-ups. Institutional change may be a prerequisite for risk reduction, as discussed further in Chapters 10 and 11;

(5) Public financing or financing with a sovereign guarantee and no risk capital, as known from Great Belt and Øresund, does not reduce risk or risk costs. It only transfers risk from lenders to taxpayers, and so is likely to increase the total risks and costs of a project.

8 Conventional megaproject development

The conventional approach

In concluding the previous chapter we explained how institutional, organisational and financial set-ups of megaprojects might significantly influence risks and costs in such projects. We therefore concluded that institutional issues and issues of risk need to be analysed together in project development. In what follows, a number of such issues are identified. In this chapter we focus on what we call the conventional approach to project development and appraisal. In the next chapter, the focus will be on approaches with a more recent history, including the so-called BOT, build-operate-transfer, approach.

Projects developed according to the conventional approach are typically financed by public money, or are backed by public (sovereign) guarantees. The majority of projects reviewed in the previous chapters were developed according to this approach, the Channel tunnel being a notable exception.[1] The conventional approach was used for the Øresund and Great Belt projects. Table VIII.i presents an outline of the steps that make up the approach, as used in these projects.

The characteristics and problems of the conventional approach to project development and appraisal are the following:[2]

(1) The project cycle does not include a pre-feasibility phase before the decision to carry out a full-scale investigation is taken. The result may be an over-commitment of resources and political prestige at an early stage;[3]
(2) Project development and appraisal are seen as technical exercises with a focus on technical solutions at an early stage. The discussion of policy objectives to be achieved by projects gives way to a discussion of technical alternatives from the beginning;
(3) Concerns related to the external effects of projects are not addressed until later in the project cycle. This may lead to project changes at a stage when such changes are particularly costly. It may also lead to the destabilisation of projects as issues surface that need to go through

Table VIII.i *Steps in the conventional approach to project development*

Steps	Action	Responsibility
1.	Identify alternatives	Government
2.	Draft terms of reference; recruit consultants for feasibility study	Government
3.	Undertake feasibility study • preliminary design and cost estimates • market analysis • economic analysis • financial analysis	Consultants
4.	Draft terms of reference; recruit consultants for evaluation of safety aspects of different alternatives	Government
5.	Carry out safety study	Consultants
6.	Draft terms of reference; recruit consultants for environmental impact study	Government
7.	Undertake project appraisal: make recommendation to government	Consultants
8.	Make decision (supplementary studies possible before final decision)	Government/Parliament
9.	Establish state-owned enterprise (SOE) to implement project	Government
10.	Application for required permits (1st phase: approval of preliminary design); preparation of documentation	SOE
11.	Mobilise finance	SOE/Government
12.	Recruit consultants to prepare detailed design and for supervision	SOE
13.	Preparation of detailed design	Consultants
14.	Application for required permits (2nd phase: approval of detailed design)	SOE
15.	Recruit contractors	SOE
16.	Supervise	Consultants
17.	Commission and initiate operations	SOE

public hearings or need formal approval by authorities at a stage when the scope for making changes is low;

(4) Negatively affected stakeholder groups and the general public are involved only to a limited extent and at a late phase of the project cycle. Input from stakeholder groups and the public is under-utilised, for instance in setting performance standards. Public dissatisfaction with projects may increase owing to the simple fact that stakeholder groups and the public are under-informed and feel left out;[4]

(5) No risk analysis is carried out;

(6) Institutional, organisational and accountability issues relating to the implementation, operations and economic regulation of proposed

projects play a minor role as part of project preparation. The problem here is lack of definition of the regulatory regime that will apply to a project and the consequences of this for risks and costs and for the overall appraisal and decision-making process.

The issue of risk analysis was dealt with in the previous chapter. Here, the focus will be on the following three institutional issues:

(i) lack of involvement of stakeholders and the general public;
(ii) lack of identification of public interest objectives; and
(iii) lack of clearly defined roles.

The following sections will successively deal with each of these issues.

> Public dissatisfaction with projects may increase because of the simple fact that stakeholder groups and the public are under-informed and feel left out.

Stakeholder involvement and public participation

Other studies that have used the Øresund project as a case in point have shown that a problem with the conventional approach to decision making for major transport infrastructure projects is that this approach tends to be characterised by close interaction between the political and government establishment on the one side, and private business on the other.[5] Citizens who are directly affected, other stakeholder groups who are concerned with the outcome of the process and the general public are not involved, or are only involved to a limited extent. Such parties also tend to receive information at a late stage, when the groups who primarily influence the decision have reached their agreement.[6] This lack of public involvement, combined with the involvement of special-interest groups who stand to benefit from the project, increases the risk of capture of the decision-making process by these interests. Politics is and should always be based on other input than expert analyses, but capture by special-interest groups often results in feasibility studies and other analyses becoming irrelevant in deciding whether or not to go ahead with a project, and in determining which alternative to build, since special and not public interest becomes the decisive factor. Thus, for both the Øresund and Great Belt projects the alternatives chosen were not the ones pointed to in feasibility studies.

The lack of public involvement also tends to generate a situation where those groups who worry about the project and are left without information and influence are inclined to act destructively, for instance by trying to shoot down the project through adverse actions inside or outside Parliament. It is here we find one of the familiar idiosyncrasies associated with the conventional approach to the development of major infrastructure projects: politics and public debate are polarised. In Sweden you were either for or against the fixed link across Øresund. In Denmark the situation was the same, for both the Øresund and the Great Belt links. At first sight – and without knowing the background – this may seem a peculiar way of viewing these projects, since tunnels and bridges are in themselves uninteresting objects. What should be of interest is the kind of result that they produce, their outcomes, and not the objects themselves. But given the fact that decisions on these major investments are essentially taken without the involvement of ordinary people – most of whom have to bear the risk of the investments – it is not surprising that many feel left out and disappointed and that this shows up in opinion polls.

If groups who feel concerned were included in the project development process for a large-scale project at an early stage, the result would be improved chances that those conditions that people view as important to making a decision would be taken into account. In our judgement, therefore, when developing and appraising megaprojects, concerned groups should be allowed to play an active role in the planning process, including a constructive role in defining the major objectives and requirements to be taken into account in the technical, environmental and economic design of possible projects.

Public interest and technical solutions

A second problem associated with the conventional approach lies in its preoccupation with technical solutions. In large-scale projects, political parties, government administrations or various lobby groups often tend to promote, or try to block, specific technical solutions. It is argued, for instance, that a tunnel and not a bridge should be built, or vice versa, or that it should be a connection for trains only and not a road and rail link, or vice versa, and so on. The pro and con positions tend to be based on only some aspects of the problem, and rarely take all features into account. This is quite natural because different groups stand to benefit from different solutions and because a fuller picture of a proposed project cannot normally be formed at an early stage of the process. But it is in this early stage that the interested and involved groups make up their minds in ways that often do not change later, even if better and more relevant information is made available.

Hence it is important to develop a planning process that is less concerned with technical solutions and information about these, and has more focus in the early stages on the requirements with respect to the economic performance, environmental sustainability and safety performance required of the project. The objective is to reach, as far as possible, consensus on these particular issues, for instance that a project will be implemented only if it is self-financing and involves minimal risk for taxpayers, if it contributes positively to achieving government policies regarding transport and the environment, and if it will result in goal-achievement in relation to national policies for traffic safety. It is not until these basic parameters of the planning process have been established that it is really meaningful to start to identify the technical solutions that would be able to meet these requirements.

It should be emphasised that the way the decisions on the fixed connections across Great Belt and Øresund have been reached is quite different from how decisions are made when committing major investment funds to new projects in the context of international development financing. The normal approach – for example as used by the major international development banks as well as in the private sector for projects of a similar nature – is thus to first identify such things as required minimum rates of return, and acceptable risk levels, but also basic environmental and safety requirements, that is the performance that must be attained by a particular project seen in a wider perspective, before possibly proceeding to preliminary or final design of the project. In our judgement such a – normal – approach should also be applied when approaching the question whether or not to go ahead with megaprojects in the public sector.

Conflicting roles for government

A third issue raised by the conventional approach is that the roles of the various parties to the project development process are not clearly defined. The conventional approach defines the role of government as being in control of every step of the process, including the choice of technology and mode of implementation, operation and financing. If the government decides that the mode of implementation and operation should be a state-owned enterprise, and that the facility should be financed by public money or with funds backed by a sovereign guarantee, government takes on the role of being a producer of commercial services and of being the financier of the venture by directly or indirectly committing the wealth of ordinary people for financing or as the collateral for the project.

In the conventional approach, government consequently plays a host of roles, some of them in essence conflicting. The question has to be asked whether a government can act effectively as both promoter of a project, and the guardian of public interest issues such as protection of the environment, safety and of the taxpayer against unnecessary financial risks. The answer is negative. There is a conflict of interest for the government here, and this conflict is well illustrated by the kinds of concern over economic, financial, environmental and safety issues that have surfaced in the wake of the decisions on the Great Belt and Øresund projects. Therefore the roles of different actors in the planning process for major transport infrastructure investments should be more clearly defined, and in particular the role of the government *vis-à-vis* other actors. Below, we return to the question of how this can be done.

> The question has to be asked whether a government can act effectively as both promoter of a project, and the guardian of public interest issues such as protection of the environment, safety and of the taxpayer against unnecessary financial risks. The answer is negative.

Institutional deficiencies of the conventional approach

Summing up, we note that in most nations and contexts, policies for how to deal with mega infrastructure projects have not been established. As a consequence, an appropriate institutional framework, namely a framework that would provide for transparency and other checks and balances to enforce accountability, does not exist for the development, planning, implementation and operation of megaprojects. In this chapter we have identified three main institutional deficiencies with the conventional approach to project development and appraisal:

(1) Under-involvement of the general public and of stakeholder groups concerned by outcomes; over-involvement of business lobby groups;
(2) Lack of identification of public interest objectives to be met by projects;
(3) Lack of clearly defined roles for government and involved parties.

In Chapters 10 and 11, we develop institutional changes that would help overcome these deficiencies. But first we will examine, in Chapter 9, what role, if any, the private sector may play in such change.

9 Lessons of privatisation

In this chapter we will review experience with private financing of infrastructure, in particular the experience of private-sector involvement in the implementation and operations of large transport infrastructure projects. The focus will be on whether private-sector involvement may help to deal with the institutional shortcomings of the conventional approach identified in Chapter 8, and thus to improve accountability, including enhancing risk management and overall project performance.

Private megaproject financing

During the nineteenth century, private capital played an important role in the development of infrastructure, in particular for investments in railways.[1] Then, in the twentieth century, public financing became much more common, including financing with private capital secured with a government guarantee. The latter type of financing has been particularly common in developing countries, where the international development financing institutions have played a major role in the financing of infrastructure investments. Such institutions normally lend against a sovereign guarantee.

During the last ten to fifteen years, there has been a resurgence of interest in private financing and also increased mobilisation of private capital for infrastructure. One reason is increased fiscal pressures on most governments with a resulting shortage of public funds.[2] Other reasons relate to the types of problem described in Chapters 2–4, for instance the large cost overruns typical for many major infrastructure projects, leading to a desire to: (i) shift the risks of projects from taxpayers to capital markets; and (ii) promote private and entrepreneurial initiative in infrastructure projects, including in the design and development of projects.[3]

An additional reason for increased private-sector involvement in the development of transport infrastructure is that such infrastructure basically produces straightforward services. Transport infrastructure is an

Lessons of privatisation

input into the production of transport services. From that starting point, in many respects a bridge or a tunnel is not fundamentally different from the investment in houses or in the building of factories. In market economies, it is normally accepted that private capital should be mobilised and that decisions on such investments should be taken by the private sector, in the sense that services will not be forthcoming if the private sector does not decide to produce them. Hence, when it comes to the production of such services, the role of the public sector is generally recognised as being to ensure that production is subjected to appropriate regulatory policies, whether they be economic to ensure efficiency in production and distribution, safety to reduce hazards, or environmental to contain or eliminate negative environmental impacts. Ultimately, such regulatory policies may mean that certain types of production activity will not be allowed, for instance because they are too dangerous or too polluting.[4]

Since the mid-1980s, governments, entrepreneurs and bankers have therefore attempted to transform so-called project finance approaches developed primarily in the oil, gas and mineral resources sector into techniques suitable for funding major infrastructure projects, including toll roads, tunnels and bridges.[5] The enterprises established to implement infrastructure projects according to this approach are often created through the award by a government agency of a concession to a private sector or mixed (public and private-sector) company. These companies have often employed a so-called build-operate-transfer (BOT) approach to the task of raising the funds needed and to implement the project (see Figure 9.1).

In a build-operate-transfer structure, entrepreneurs (usually major construction contractors, equipment suppliers, property developers and experienced public utility operators) form consortia, often with other investors, to make substantial equity investments in a special-purpose project company. The project company then builds the required facility, operates it long enough to pay back the debt finance raised and remunerate shareholders, and eventually transfers title, rights and responsibilities of further operations of the facility to the government. In situations where the project company also owns the facility throughout the period, such a structure is referred to as build-own-operate-transfer (BOOT).[6]

The decision by private-sector investors to fund a major infrastructure project to be implemented and operated in terms of a concession is based on a different form of analysis from a customary investment decision in an ordinary company in the private sector. This is for the following reasons:
(1) The assets of infrastructure projects often have little or no resale value. If the right to collect fees for making use of the infrastructure is lost, the residual value of the project company will be small. In customary

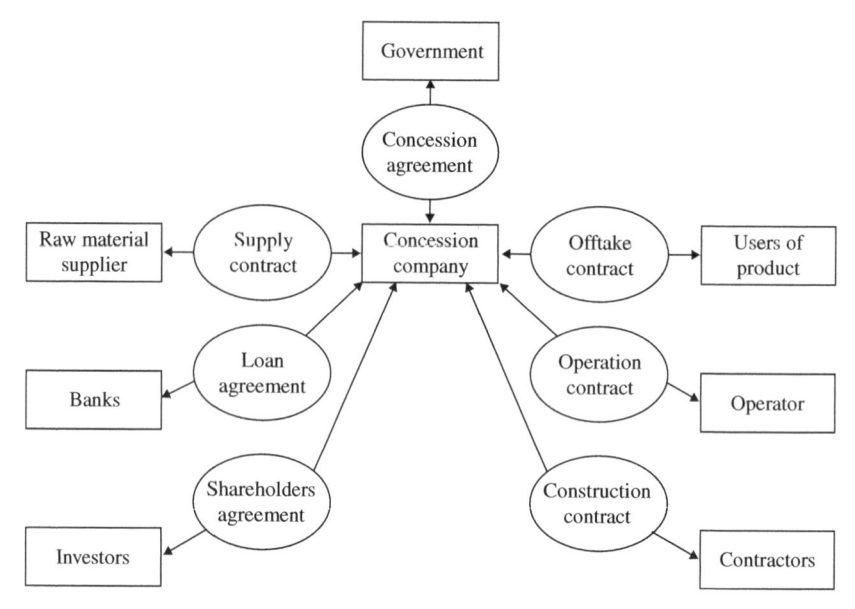

Figure 9.1 The typical BOT (build-operate-transfer) approach to megaproject development
Source: S. C. McCarthy and R. L. K. Tiong, 'Financial and Contractual Aspects of Build-Operate-Transfer Projects', *International Journal of Project Management*, vol. 9, no. 4, November 1991, p. 223.

investment opportunities, by contrast, asset values are significant. This is why the term 'sunk costs' is used to describe costs spent on infrastructure projects;
(2) A project company for an infrastructure project, for instance a bridge or a tunnel, often has no track record of profitability and no immediate earnings prospects. Track record, together with an understanding of expected future market conditions, normally forms the basis of an investment appraisal with respect to ordinary companies;
(3) No market exists for sale of the company itself (as it has not yet been established). By contrast, the information conveyed by market transactions in other companies, including in their shares, can be used to validate investment appraisals of such companies' worth based on assets and/or profitability;
(4) The commercial risk is different from conventional investment opportunities in competitive sectors.
At the heart of project financing is a contract that allocates risks associated with the project and defines claims on rewards. Normally, project sponsors transfer construction risks to specialised construction companies

Lessons of privatisation   95

through lump-sum contracts. According to the World Bank, the evidence on private construction, although still limited, is favourable and reflects tight contractual conditions and severe penalties for cost and time overruns. A preliminary review of the International Finance Corporation's infrastructure projects shows that time overruns in construction have been seven months on average and cost performance has been about on target.[7]

A characteristic of the financing approach for build-operate-transfer schemes is that the transactions are highly complex, are difficult for experts outside the financial sector to understand in detail, and are based on lengthy contractual relations between the different parties involved in the project. Many attempts at arranging 'bankable' concession packages have failed to meet the requirements of the financial market. However, provided a concession can offer some operating cash flow, that is tolls less operating cost and tax, it should normally be possible to attract private investment.[8] There have been a number of privately financed infrastructure projects where the expected value of operating surpluses has greatly exceeded the investment costs, for example the concession to build and operate new container terminals in Hong Kong. In these cases, governments have even been able to extract substantial licence fees from operators as a result of fierce bidding in the course of a tender process for the concession.

In the case of toll roads the reverse is often the case, and in such cases government will need to make up the difference between the private capital injection and the total investment cost, if the roads are to be built.[9] Typically, this has been done by providing land for free, or on the basis of deferred payments, namely by sharing or dedicating toll revenues from other roads (for example Bangkok Second Stage Expressway, Sydney Harbour Tunnel, Dartford Bridge), or by direct grants or subsidies.[10] The Manchester Light Rail project in the UK followed the last route to its logical conclusion in awarding the concession to the bidder requiring the least level of government grant. A similar approach was employed to finance the Arlanda Airport rail link in Stockholm.

Lessons of the privately funded Skye Bridge project, UK:
1. Private-sector finance may be hard to mobilise because of the perceived levels of risk and uncertainty;
2. Allowing full inflation adjustment of the toll charge may provide considerable potential windfall gains to the concessionaire;
3. Only projects offering the highest return are likely to attract potential interest.[11]

The case of Eurotunnel

The Channel tunnel is the 'flagship' of concession financing in the transport sector. It is a build-own-operate-transfer (BOOT) project with a concession, originally lasting fifty-five years, later extended to sixty-five years as part of a settlement regarding responsibility for cost overruns, and later again extended to ninety-nine years in an attempt to secure project viability (see Chapter 4).[12] The project has been through many ups and downs, including a suspension of interest payments on loans and a fire that damaged image and revenues. Whether the project will eventually become a success or a financial failure, only time will tell.

The project has come to cost about twice as much as originally envisaged. Part of this increase can be ascribed to enhanced safety, security and environmental demands, highlighting the policy type of risk that is associated with this kind of project. For instance, production of rolling stock was halted to make possible a 10 cm increase in the width of all pass doors. This cost Eurotunnel around £40 million and, more seriously, months of delay and additional capital costs. These additional costs had to be covered by Eurotunnel, because the governments had successfully insulated themselves from any financial risk and involvement. Furthermore, access links, such as a high-speed rail link from London to Folkstone which was foreseen in the near future, have still not been built. As a consequence timesavings for users are smaller than expected in the first forecasts, resulting in a lower demand figure. Again, as such types of policy risk had not been insured by the contract they were allocated to Eurotunnel. But part of the explanation of Eurotunnel's ballooning budget is also to be found in simple construction cost overruns and in the absence of a clear owner of the project company at the outset. The concession for the fixed link across the Channel was thus originally won by a consortium of contractors and financiers in 1986, whose prime interest was to win construction contracts and provide financing rather than to serve as the concessionaire.[13] As the promoting group was awarded the concession following a competition with three rival consortia there was considerable doubt with all consortia whether they would recover the cost of the bid. In these circumstances little work was done on detailed design, and cost estimates were more or less rough guesswork.

After the concession had been awarded, and Eurotunnel became the concessionaire, the company was in the beginning weakly staffed, and the staff moreover consisted almost exclusively of secondees, who were then expected to negotiate with the organisations from which they had been seconded. It seems clear that this worked to the disadvantage of Eurotunnel. There were insufficient resources at the design and development stage for a speedy development, and as a whole inadequate financial resources were mobilised at the beginning. It also appears that the contracts signed by Eurotunnel were much less tight than would be normal for a project-financing venture of this size.[14]

Eventually, however, with a change in the ownership of Eurotunnel through the selling of shares and with the appointment of new management and staff this situation was turned around. Since the late 1980s and until completion, Eurotunnel took a strong line with contractors in arguing for its rights. And in spite of the handicaps created at the outset, Eurotunnel was able to complete the tunnel basically on time. It is probable that the tough line taken by Eurotunnel in the contractual negotiations was critical to its continued survival; with a more soft approach, costs might have been higher and revenues lower than turned out to be the case, which would have made the financial crisis and restructuring of 1995–98 deeper and more difficult than it was.

The lessons learnt to date from Eurotunnel and similar projects are primarily three. First, behind proposals for concessions there may be both short-term (construction) and long-term (operations) interests, which are likely to diverge. It is important to ensure that the long-term interests play a leading role from the beginning and that this is reflected in the institutional (ownership and management) structure of the concessionaire, as will be borne out by further examples below. Various mechanisms are today being employed in connection with build-operate-transfer and similar projects to ensure that this is being achieved, including that original equity holders of the consortium winning a build-operate-transfer contract cannot sell their equity within a certain period of time without permission from government, and that bids submitted will also contain explicit commitments with regard to construction times and costs.[15] An additional mechanism developed in the UK in connection with the introduction of DBFO-contracting (design-build-finance-operate) is the right of the government to terminate the agreement with the concessionaire under certain circumstances plus the so-called Direct Agreement between lenders and the government, which allows the funders 'step-in rights' in case the contract with the concessionaire is terminated. This right thus enables the funders to take over operations themselves for a limited period of time during which they may try to find a replacement operator.[16]

Secondly, provided that a project has been appropriately conceived, developed and designed, including the concessionaire having an adequate ownership structure, build-operate-transfer structures may be efficient mechanisms for implementing projects of this nature.[17]

Thirdly, it is reasonable to identify the type of risk from the beginning and allocate it properly to the parties involved. In particular, political risk cannot be carried and managed by private investors but has to be allocated to those responsible for politics, namely the state.

Toll roads and similar projects

Whilst tolling of roads was commonly used in the eighteenth and nineteenth centuries in the USA, today toll roads are a global phenomenon with European countries taking the lead.[18] France and Spain have had the longest and most extensive experience of building private roads financed by tolls.[19] The Spanish private concessions began in the 1960s, while their French counterparts date from the early 1970s. The Italian experience also dates back to the 1960s, but the role of private operators has been quite limited there.

By the end of the 1990s, France had opened 6,700 km of tolled intercity motorways, which were operated as concessions by nine different companies. However, only one of them, Cofiroute, is privately owned. The other companies are mixed investment companies, in which both private and public ownership can be combined, but all of these companies are almost entirely owned by the national and local governments. Private involvement in the French motorway system was encouraged from the early 1970s, and several private consortia entered the market. Basic rules for the institutional set-up regulating ventures were that equity and reserves committed by shareholders had to be at least 10 per cent of the cost, and that loans without a sovereign guarantee had to be at least 15 per cent. Tolls could be set freely by private concessionaires for an initial period of time.[20]

However, all but one of the private concessions later ran into financial problems, mainly on account of the energy crises in the 1970s, and were either absorbed or converted into mixed companies during the early 1980s. At this time, private involvement in the sector was not encouraged any longer. In addition, the authorities started to intervene more heavily in the setting of tolls through its policy of toll harmonisation across the country. The problems of the private companies can also be explained in part by the fact that they were mainly owned by construction companies whose initial interest had been the construction of the motorways rather than their operation. Since the late 1980s, when French policy was changed again to support private-sector involvement, it has been difficult to mobilise private interest on account of a continued policy of tight control of tolls by the Ministry of Finance.

By 1998, there were approximately 2,250 km of motorway in Spain, which were operated by fifteen different concessionaire companies. Twelve began as private concessions, and three of these were taken over by the national or regional governments in the 1980s. As in France, the main reasons for the poor financial performance of private operators were the energy crises and under-capitalisation of projects. Again it appears to have been a problem that the major actors behind the concessionaires were construction companies that had been mainly interested in building, and not operating, the projects. These problems seem to have been exacerbated by the fact that only 10 per cent of the construction costs had to be financed from equity and that government would guarantee up to 75 per cent of foreign loans; moreover government assumed the full exchange rate risk of foreign loans.

Both France's 6,700 km and Spain's 2,250 km of toll roads appeared to be largely self-supporting in the late 1990s. France's experience suggests that private companies probably can build roads more cheaply than public companies. However, experience in both France and Spain points to the danger of creating incentives under which private concessionaires are more interested in building than operating a toll road, and to the dilemmas of regulating toll rates. The French approach to price regulation creates a policy type of risk, as the criteria are not explicit. In Spain an explicit regulatory model is used, but it is applied universally, thereby neglecting important differences between concessions.

Since these earlier precursors, the building of toll roads with private involvement has proliferated. One source identifies 121 infrastructure projects with private capital in developed countries between January 1985 and October 1998, the majority of which was toll roads. The average project value was around US$750 million, driven by a number of EU-sponsored megaprojects in Europe. According to the World Bank, 280 road projects were initiated or ongoing in developing countries during the period 1990 to 1997, involving public–private partnership. The average project size was around US$190 million, with the bulk of these projects to be found in East Asia and Latin America.[21] The People's Republic of China is the country with the most aggressive programme of all. It is estimated that 2,800 km of expressways have passed into private hands in the last five to six years in China. There are about 800 concession agreements, all of which are based on either leasing or build-operate-transfer arrangements.[22]

Major toll road developments can also be found in Mexico, Malaysia, Indonesia and California. Mexico has added 5,500 km of new toll roads at a cost of US$10 billion over a period of five years. Malaysia has built one of the most expensive public–private projects in the developing world, the US$1.6 billion North–South Expressway.[23] In California, four toll roads worth in total US$2.5 billion have been added, after an agreement was reached in January 1991, providing for a build-operate-transfer approach and without any state or federal funds being supplied.[24]

Several of these projects have run into difficulties, including the Second Stage Expressway in Bangkok and the Mexican toll road programme. In the latter case, the problems reflect inadequate preparation, cost overruns and optimistic traffic forecasts. Traffic after opening of the new roads was also lower than expected on account of competition from parallel routes. As concerns the Thai example, the source of the problem is to be found in poorly formulated regulatory policies, which, *inter alia*, meant that it was unclear who would actually collect the tolls when the project was completed.[25]

A significant lesson from private-sector involvement is that project preparation must be comprehensive and must be allowed to take the required time in order to clearly identify all issues, risks and policy concerns. The European Bank for Reconstruction and Development seemed to have learned this as well as other lessons from earlier projects as reflected in the arrangements underpinning its first project financing scheme in the road sector of Eastern Europe, which is the M1–M15 toll motorway, a build-own-operate-transfer project in Hungary. The concession company is a special-purpose company, which was required to be adequately staffed and managed to be able to draw on the credits. Lump-sum contracts were awarded for both construction and maintenance. Non-commercial risks were to be borne by the government, and the lenders assumed responsibility for managing interest rate and currency risks.[26]

In spite of such careful preparatory work, the M1–M15 project has had problems because actual traffic fell far below expectations, and the project is therefore technically bankrupt as are many other toll road projects. The main reason appears to be optimistic assumptions about traffic growth, but also that many drivers have elected to bypass the toll road by using alternative non-tolled roads.[27]

A second feature of toll roads is thus that they are risky projects, mainly on account of a significant market risk, which is more pronounced than for many other types of commercial operation. There are several reasons for this. First, the demand for road transport is very dependent on economic development. It is not unusual for road traffic to grow at twice the rate of the economy, implying an income elasticity of about two. This tendency is stronger for major roads than for smaller ones. Secondly, the price elasticity with respect to tolls is often very high; experience suggests that the elasticity typically is in the range −1.4 to −2.5, so that increased tolls will actually result in lower revenues. A major reason for this is that there are often alternatives to the toll roads; indeed it is often a requirement that there be an untolled road parallel to the tolled one.[28] Both of these problems can, as noted, be identified behind the M1–M15 project in Hungary as well as in projects in many other countries.

The significance of the market risk is exacerbated by the cost structure of motorways, with a significant portion of constant costs. A decrease in

Table IX.i *Cost and traffic development in four privately owned transport infrastructure projects. The table includes private projects for which data were available*

Project	Construction cost overrun, fixed prices (%)	Actual traffic as percentage of forecast traffic, opening year
Channel tunnel, UK, France	80	18
Third Dartford Crossing, UK	20	115
Second Severn Crossing, UK	20	100
Pont de Normandie, France	15	120

Source: Mette K. Skamris, 'Economic Appraisal of Large-Scale Transport Infrastructure Investments'.

traffic consequently does not lead to equivalent reductions in costs. It is estimated that the traffic density has to amount to at least 10,000 vehicles per day, and preferably 15,000 vehicles, for a toll road to become viable. About 1,500 vehicles are required to pay for the toll collection system, and another 3,500 vehicles for other operational costs, and the balance, that is 5,000–10,000 vehicles per day, is required to finance the capital, implying that about two-thirds of the cost is constant in the long run. In the short run, an even larger portion is constant.[29]

These problems are magnified by the condition that many toll road concessions today make up small parts of the motorway system, preventing positive cash flows made in older parts of the system to be offset against losses on more recent parts. Such cross-subsidisation is a feature of the French system, which currently provides for transfers between different parts of the toll roads by way of deposits of 'excess profits' into certain joint funds. The most exceptional solution in this regard is offered by Japan, where there is only one company – Japan Highway Public Corporation – operating the inter-city toll road system, altogether about 9,000 km. However, this company is a state-owned enterprise.

It is important to note that the issue of lower-than-expected revenues on toll roads does not necessarily mean that incorrect overall forecasts were made as part of project preparation; rather the issue is that there is a tendency to underestimate the implications of the volatility of demand over time and the implications of competing alternatives. Where there are no real alternatives available, traffic forecasts tend to be much more accurate, as suggested by the data in Table IX.i referring to three bridge projects. In all three cases the new bridge offered a significant improvement in comparison with the old solution in terms of time, a time saving that the motorists were apparently prepared to pay for.

Design and build

A way of eliminating the issues raised by toll financing of private operations of a road is for the government in lieu to ask the private operator to provide all the services required but then for the government to remunerate the operator directly for these. In other words, the private sector provides for everything, including all the required working capital, but the services are ultimately paid for by the government by way of appropriations. This kind of public–private partnership, which has recently been introduced, can be expected to grow significantly in the future. The approach has so far mainly been used for contracting for maintenance and rehabilitation activities in the road sector, for example in Argentina, Australia and New Zealand, and is also in the process of being introduced in several other countries in Latin America and Africa.[30]

The best-known examples of this approach are the DBFO projects (design-build-finance-operate) in the UK, which are investment projects. In terms of the DBFO approach, a contract to design, build, finance and operate a new road is awarded to a concessionaire – a special-purpose company – which is then remunerated directly by the client, the state, over the concession period. In the case of the UK, where 9 such contracts were awarded during the 1990s and a further 4 were under procurement in 2000, these contracts have a duration of 30 years.[31] Maintenance and rehabilitation contracts have so far been for a much shorter duration, namely 5 to 10 years.

Investment projects based on the DBFO approach are also found in other countries, such as Holland, Finland and Portugal, and are being employed in railway infrastructure projects as well. An approach of a similar nature has been used in Germany, referred to as *Vorfinanzierung* (prefinancing). In Germany, twelve road projects involving DM 4.6 billion were financed in this way during the 1990s, as well as one railway infrastructure project. These contracts, however, do not entail design and operations upon completion, implying a significant difference between this approach and that of the projects described below.[32] One of the most advanced German projects is the Warnow tunnel in Rostock, which links two city districts. A concession has been given to the French Bouygues company. After awarding the concession, substantial problems emerged because a build-operate-transfer concept includes user charges to finance the project. But forecasts had been made without considering such charges and the forecasts thus had to be corrected. After correction the project is no longer viable and the EU Commission and the German Federal Government have been asked for financial aid to save it. As these institutions are not responsible for urban transport investments it is still unclear whether the barriers to the Warnow crossing tunnel can be removed. This example shows again the type of difficulty that is bound to occur for public–private partnerships if the roles of the public and of private investors are not clearly defined from the beginning.

A feature of the DBFO projects in the UK is that remuneration is mainly based on shadow tolls. One stated purpose of this is to transfer the traffic – or the market – risk to the private sector, thereby reducing the overall cost of risk. It may, however, be shown that it is not possible to transfer the traffic risk by way of shadow tolls and that the actual shadow tolling structure employed so far in the UK really bears this out.[33] Indeed, in the most recent DBFO projects, shadow tolling is playing less of a role, and a remuneration system based on accessibility is being put in its place. This also applies to a similar system being developed in Sweden as well as to rehabilitation and maintenance projects in other parts of the world, where the remuneration is based on a fixed fee per month.[34]

An important feature of these more recent developments in the road sector is that a new role is emerging for the private sector in infrastructure provision. The focus has moved from financing towards cost-efficiency in building and maintaining roads, and towards innovation in road-building techniques. What DBFO thus has brought to the fore is the importance of this approach to contracting in infrastructure provision.

Traditionally, contracting in the building sector has been based on inputs. The client has been responsible for deciding on the techniques and materials to be used and the contractor for delivering according to set specifications. This approach has hampered innovation but is typically used all over the world today. Slowly, however, this way of doing things is being challenged by contracting based on performance specifications. Such contracting covers construction *and* operations, in order to verify performance, and requires clients to formulate their demands in performance terms, leaving design, techniques and materials to be the responsibility of contractors. One purpose is to allow contractors to invest in innovations by not requiring disclosure of the techniques and materials used until expiry of the contract, effectively giving the contractor a patent period. In essence, this was also the intention of the UK DBFO contracts mentioned above, although the ambitions have only partially been realised in this regard. Further developments in the DBFO approach, however, are likely to lead to more performance-based contracting.[35]

In sum, experience clearly varies for build-operate-transfer and related schemes. At one extreme, experience from Hong Kong has proven the benefits of such schemes. At the other extreme, experience from nearby Thailand has shown just how tricky and risky build-operate-transfer schemes may be if not well thought through. The main ingredients in a successful build-operate-transfer project seem to be an experienced and simple governing body and structure, an intact contractual agreement, a structured set of regulations, a large and reliable consortium, an experienced construction organisation, an uncorrupted political regime and no intervention from politics once the contractual agreement has been decided.[36]

> While far from offering a panacea to the risk and accountability problems for megaprojects, given an appropriate and properly implemented institutional framework, private involvement may be helpful.

Privatisation on balance

The point of departure for this chapter was the question whether private involvement in major infrastructure projects may help to increase accountability and thus protect the public better than projects without private involvement. The review of recent experience with private-sector involvement shows that the answer to this question is not straightforward. While far from offering a panacea to the risk and accountability problems identified for the conventional approach to project development, given an appropriate and properly implemented institutional framework, private involvement may be helpful in certain ways, although the experience of private-sector involvement with transport infrastructure is still too limited to allow firm conclusions. In our analysis the preliminary lessons learnt are as follows.

First, private-sector involvement requires a well-prepared, capable and committed client (government agency) with a clear vision and understanding of the role to be played by the private sector. Far too many projects with private financing fail to achieve what is being hoped for on account of the project being inadequately prepared, for example because it is 'driven' by private or narrow political interests.

Secondly, private-sector involvement requires a long-term commitment on the part of the owners of the involved private-sector companies. This is a chicken and egg situation, because appropriately structured private-sector companies will not be forthcoming until there is a large and sustained volume of projects requiring the involvement of the private sector.

Thirdly, the market risk associated with transport infrastructure projects is substantial. The primary reason for this is that the demand for transport is highly income-elastic. Thus, if the economy of a country is growing rapidly, the traffic will typically grow even quicker and consequently also toll revenues. In view of the low level of the variable costs of infrastructure, the impact on overall financial performance will be significant. On the other hand, if the economy is sluggish, the performance of a toll road will often be even poorer. A private-sector company can do little to influence this situation. In addition, tolled facilities are often subjected to competition from alternative routes, including parallel 'free roads', putting limitations on the use of tolls as an instrument to increase revenues. The demand elasticity of tolls is often high, making it difficult for a company to increase revenues by raising toll levels.

Fourthly, the market risk from the point of view of the toll company is magnified by the presence of regulation of the toll levels, directly or indirectly. The presence of regulation is often motivated by a desire to ensure that the private-sector company is not making monopolistic profits, or the official motivation may be to ensure that the tolls do not restrict use of the facility too much, resulting in an under-use from an economic point of view. A problem of private-sector involvement in transport infrastructure is that it may come into conflict with the objective of economic efficiency.

A fifth lesson learnt is that the best use of the private sector is to control construction and operating costs and to foster the development of new techniques for the construction and operations of infrastructure. A precondition for this is, however, that contracting with the private sector is done on the basis of output (performance) specifications, and not input specifications.

In sum, private-sector involvement in the provision of transport infrastructure may help identify risks more clearly, reduce risks and place risks with those best able to bear and manage them. Private involvement will typically also increase pressures for performance. Finally, private involvement may help mitigate problems with lack of clearly defined roles for those involved in major infrastructure projects and problems with rent-seeking behaviour from business lobby groups.

By way of concluding this chapter, it should be added that the problems identified for the conventional approach with under-involvement of the general public and with lack of identification of public interest objectives to be met by a given project will have to be solved in other ways than by private involvement. Hence, another and final lesson from the review of experience with private-sector involvement in major infrastructure projects is that public–private collaboration is crucial, even for private-sector projects. A review of experience from the Channel tunnel puts it this way: 'Perhaps the most important lesson to be drawn, even from private sector projects like the Tunnel, is that public–private sector partnership remains essential.'[37]

> Public–private collaboration is crucial, even for private-sector projects.

10 Four instruments of accountability

The main shortcoming of the conventional approach to appraisal and development of megaprojects is the lack of mechanisms for enforcing accountability, namely the absence of, on the one hand, clear objectives and, on the other, arrangements for: (i) measuring how objectives are being met; and (ii) rewarding good and penalising poor performance. In this chapter, we will identify a number of basic instruments that we see as necessary for strengthening accountability. However, before discussing these instruments, we consider in more general terms the roles to be played by the private and public sectors in the development of megaprojects. Understanding the appropriate roles of the two sectors is fundamental to the identification of a process for appraisal and decision making that will ultimately work in the public interest.

Redrawing the borderlines of public and private

A basic issue that is both complex and likely to stir controversy is the question whether megaprojects should be publicly or privately led. Neither the complexity of the issue nor its potential for controversy should come as a surprise. After all, one of the most fundamental aspects of public policy is at play, namely that of defining the frontier between the public and private sectors.[1]

For major infrastructure projects one could argue an either/or position on public *versus* private leadership:
- either such projects should be placed entirely within the public sector – for example in a government department, an agency or a state-owned enterprise – to ensure accountability through the rules of transparency and public control that apply to the public sector;
- or such projects should be placed entirely in the private sector – for example by means of build-operate-transfer or other concession arrangement – to ensure accountability through competition and market control.

Such an either/or position may have pedagogical merit. In practice, however, it is an untenable oversimplification, if for no other reason than simply because there is no such thing as an entirely private venture for investments of the magnitude and consequences considered here. For instance, the mere task of communicating, monitoring and enforcing laws and regulations regarding safety, environment, economy, and so on requires extensive public–private collaboration, even for the most private of projects, not to speak of possible concessional issues.

Therefore, the point of departure for deciding on the division of work between the public and the private sectors must be a recognition of the fact that the risks for poor performance in planning, construction and operations of investments of the size and complexity at stake are so high, and the consequences so pervasive, that every conceivable means to introduce checks and balances on performance must be considered, be they public or private. Thus public-sector measures need to be combined with private-sector ones in ways that maximise overall accountability and performance. Before looking at how this can be done, it is instructive to look briefly, once again, at the experience from the multibillion-dollar bridge and tunnel projects at Great Belt and Øresund.

Public-sector involvement should be strengthened by government taking a more active role in:
- Engaging stakeholders and the public
- Identifying public interest objectives
- Defining regulatory regimes

As mentioned, in the public sector the main mechanism to enforce accountability is transparency, and in the private sector it is competition. The Great Belt and Øresund projects have been set up institutionally as public joint-stock companies with full state ownership and financing backed by sovereign guarantees. We see three main problems with this institutional arrangement regarding accountability:

(1) By placing the Great Belt and Øresund projects in joint-stock companies, transparency, and with it the possibility for political and public control, has been reduced substantially. This is because in Denmark, the Freedom of Information Act and similar measures that guarantee public scrutiny of government activities do not apply to this type of company;[2]

(2) By choosing state ownership of all shares, and hence eliminating the sale of shares to private investors, shares from the Great Belt and Øresund companies do not compete on the stock exchange with shares from other companies. Therefore, no price mechanism and no stock market reviews exist to indicate how the projects are faring, with the transparency and pressures for performance that this would entail;

(3) Finally, the Great Belt and Øresund projects are mainly financed with loans that are secured with sovereign guarantees. These are loans for which the lenders have little incentive to monitor and put pressure on performance, since payback is secured through the government's power to tax.

Public-sector involvement should be weakened in the following ways:
- No total sovereign guarantee should be given to lenders
- Government should not act as project promoter, but should, instead, enforce the arm's-length principle

---

Summing up, the Great Belt and Øresund projects lack the transparency and public control that placement in the public sector proper would entail. The projects also lack the pressure on performance and risk reduction that placement in the private sector would entail. In short, as regards accountability and performance, the Great Belt and Øresund projects 'fall between two chairs', as the Danish proverb has it. A study from the Danish Ministry of Finance singles out the Great Belt and Øresund projects as liable to a 'risk of lack of efficiency' owing to 'lack of sufficient market pressure'.[3]

In contrast with the role of the public and private sectors in the Great Belt and Øresund projects, the following basic institutional rearrangement of public and private responsibilities is suggested for major infrastructure investments in order to strengthen accountability. Other adjustments may be relevant, depending on the regulatory regime chosen:

(1) Public-sector involvement should be strengthened by government taking a more active role in:
- Engaging stakeholder groups and the general public in planning and decision making from an early stage;
- The identification, to the extent possible up front, of public interest objectives and requirements that the project must meet;

- Defining, to the extent possible up front, the regulatory regime that will apply to the project, if implemented, including the principles of public and private involvement.

(2) Public-sector involvement should be weakened in the following ways:
- No total sovereign guarantee should be given to lenders, only a limited guarantee, if any;
- Government should not see its primary role as that of project promoter, but should, instead, keep the project, and involved actors, at arm's length in order to critically assess, at all stages, whether the project meets public interest objectives and requirements, and complies with laws and regulations, for instance regarding environment, safety and economy.

(3) Private-sector involvement should be strengthened in the following ways:
- Some degree of private risk capital should be used to finance the project;
- Private consortia bidding for construction and operations contracts should have more of an opportunity to suggest which technical solutions and designs will best meet the public interest objectives and requirements identified by government.

(4) Private-sector involvement should be weakened in the following way:
- Business lobby groups, and other special-interest lobby groups, should be given less opportunity for rent-seeking behaviour.

Private involvement should be strengthened by:
- Involving a degree of private risk capital
- Involving private consortia in performance-based project design

---

Basic instruments of accountability

Having redrawn the borderlines for the roles of the private and public sectors, it is now possible to identify four basic instruments that are key to the establishment of an appropriate process and institutional set-up for the development of major infrastructure projects:[4]

(1) Transparency;
(2) Performance specifications;
(3) Explicit formulation of the regulatory regime, and clear identification – and where relevant, elimination – of policy risks before decisions are taken;
(4) The involvement of risk capital.

Below, each of the four instruments of accountability is described in more detail.

> Private involvement should be weakened:
> • Lobby groups should be given less opportunity for rent-seeking behaviour

Transparency

The acid test of public scrutiny is the main means of enforcing accountability in the public sector. The role of government is, in principle, to represent and protect the public interest (as defined by Parliament or legal precedent) and therefore it must at all times be possible for the public to verify whether this is indeed the case. The transparency requirement means, *inter alia*, that all documents and other information prepared by the government and its agencies should be made available to the public.[5]

Since major infrastructure projects are among the most costly ventures in a society, and since the ordinary citizen as taxpayer is often the ultimate guarantor for such projects, it is hard to find legitimate reasons for not informing citizens fully about projects, and for not letting citizens have a say concerning what they think about them. Consequently, two-way communication with civil society, and with stakeholder groups and media, should be given high priority. The task of communication and participation should be taken as seriously, and should be funded as adequately, as the technical, environmental and economic tasks in a project, right from the early planning stages. Professional expertise on communication and participation should be used, not for PR purposes but in order to ensure that communication and participation will be effective and contribute to all parties being heard, just as professional expertise is used for most other tasks in major project development.[6]

Stakeholder groups and civil society should be invited to participate from an early stage and throughout feasibility studies and decision making.[7] Who, exactly, should participate will vary from project to project. But in principle no party affected by a given decision should be excluded and all participants should have equal possibility to present and criticise validity claims. Participation should be as representative as possible. This is best ensured by government, or parties acting on behalf of government, taking an active role in identifying, inviting and balancing stakeholder and civil society groups, so that all relevant groups get an opportunity to participate, and no one group gets to capture and dominate this aspect of the process.[8] All documents prepared or commissioned by the government should be released to groups and to the general public as they are produced. Feedback from participating groups should be actively used in the feasibility studies and in the decision-making process, including a constructive role for the groups in defining the major requirements to be taken into account in the technical, environmental and economic design of possible projects.

Public hearings, social surveys, arbitration, advisory committees and other means to communicate with a wider public should be considered and used from an early stage, for the same reasons and purposes that stakeholder and civil society groups should be identified and involved.[9] In addition to these traditional and well-tried methods of citizen participation, more recent models should also be considered such as citizen panels, citizen juries, citizen advisory committees, citizen initiatives, compensation and benefit sharing, negotiated rule making, mediation and Dutch study groups, as proposed by, for instance, Ortwin Renn, Thomas Webler and Peter Wiedemann.[10] Research shows that although traditional models of participation are useful in many instances, they are wanting in others with respect to both quality of decisions and democratic distribution of power. Ongoing development and tests of new participatory models are therefore necessary to improve transparency and accountability.

Given the amount and complexity of information produced as part of the feasibility study for any megaproject, and given that the track-record of megaprojects shows such information to be often unreliable and biased (see Chapters 2–6), it will in many instances be difficult to decide whether information produced by project promoters and their consultants is state-of-the-art and balanced. To assist in deciding this question, peer review should be established, both as traditional peer review and as what could be called 'extended peer review' carried out by 'extended peer communities'. Silvio Funtowicz and Jerome Ravetz define extended peer communities in the following manner:

[T]here must be present an expertise whose roots and affiliations lie outside that of those involved in creating or officially regulating the issue [here megaprojects]. These new participants, enriching the traditional peer communities and creating what might be called *extended peer communities*, are necessary for the transmission of skills and for quality assurance of results (italics in the original).[11]

In answering the question 'who are the peers?', Funtowicz and Ravetz include the traditional peers, that is, scientists and experts who are colleagues working within the 'paradigm of the official expertise'. Funtowicz and Ravetz go on to stress, however, that with extended peer communities

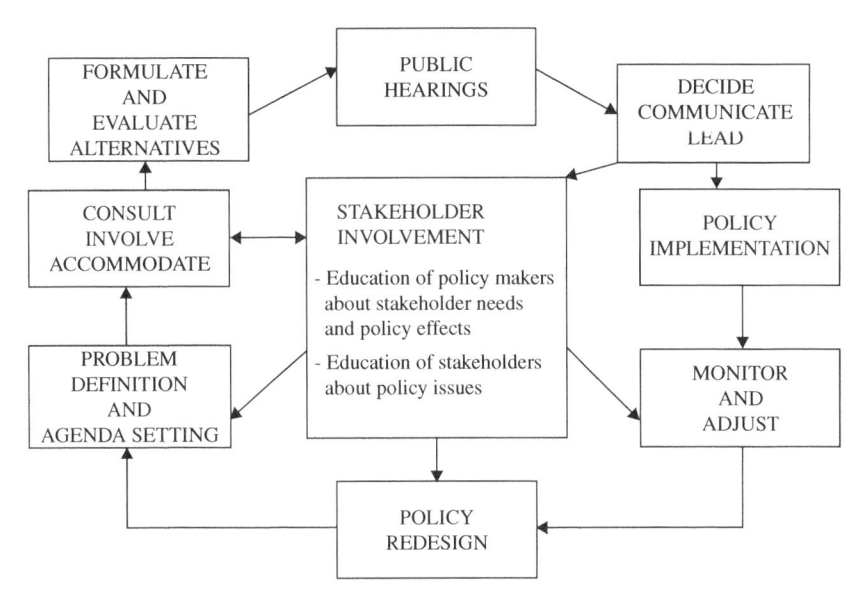

Figure 10.1 A stakeholder-based approach to decision making
Source: Adapted from John A. Altman and Ed Petkus, Jr, 'Towards a Stakeholder-Based Policy Process: An Application of the Social Marketing Perspective to Environmental Policy Development', *Policy Sciences*, vol. 27, no. 1, 1994, p. 39.

the traditional peers are enriched 'at the very least' by the contribution of other scientists and experts, who are 'technically competent but representing interests outside the social paradigm of the official expertise'. Finally, Funtowicz and Ravetz emphasise that quality assessment cannot be left to the experts themselves. It must also include '[t]hose whose lives and livelihood depend on the solution'. Such people are seen, in effect, as a third kind of expert. Thus the relevant publics are no longer merely 'impacted', they are knowledge generators.

Traditional peer review of environmental issues was carried out with success for the Great Belt and Øresund projects (see Chapter 5). For the planned Baltic Sea link between Germany and Denmark at Fehmarn Belt – another transnational multibillion-dollar infrastructure project which has been given priority by the EU Commission as part of the Trans-European Transport Network – elements of extended peer review have been introduced in feasibility studies in order to evaluate cost estimates, traffic forecasts and environmental impact assessments produced by project promoters and their consultants (see Appendix). The experience from these and other projects shows that both quality of results

114   Megaprojects and Risk

**Organisation of the first phase of debate on
the main functions of the infrastructure**

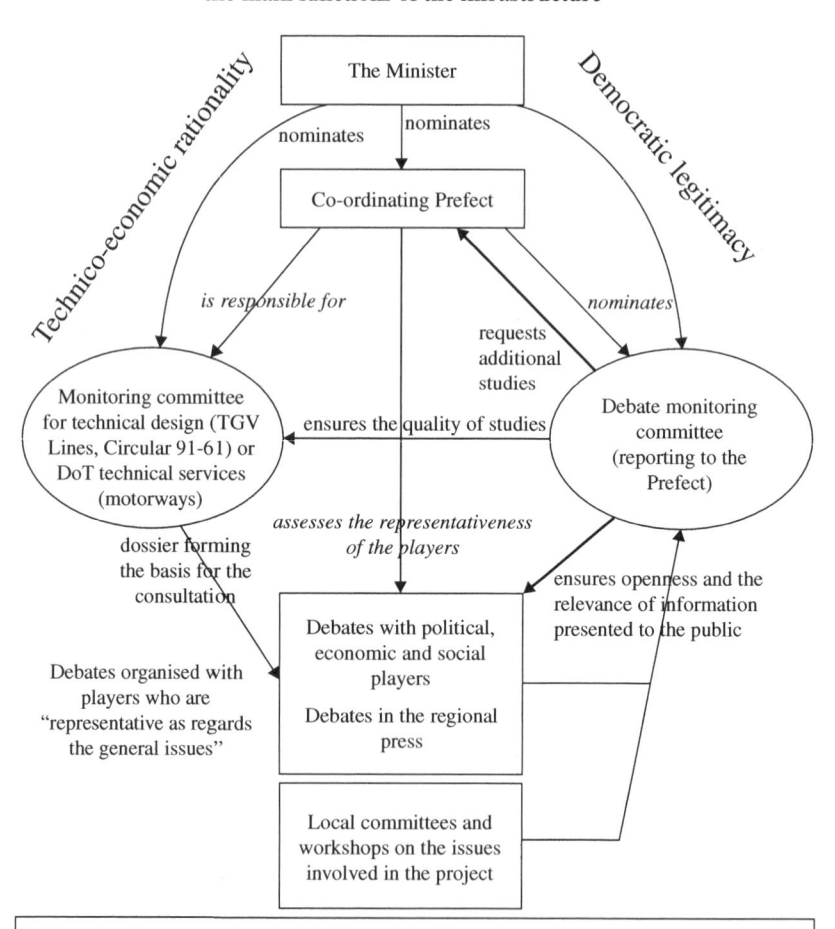

Technico-economic rationality *Democratic legitimacy*

The Minister
— nominates → Co-ordinating Prefect
— nominates (left) / nominates (right)

Co-ordinating Prefect:
- *is responsible for* → Monitoring committee for technical design (TGV Lines, Circular 91-61) or DoT technical services (motorways)
- *requests additional studies* ← Debate monitoring committee (reporting to the Prefect); *nominates*
- *ensures the quality of studies* (from Debate monitoring committee to Monitoring committee)
- *assesses the representativeness of the players*

Monitoring committee → dossier forming the basis for the consultation

Debates organised with players who are "representative as regards the general issues"

Debates with political, economic and social players
Debates in the regional press

Debate monitoring committee ensures openness and the relevance of information presented to the public

Local committees and workshops on the issues involved in the project

---

The co-ordinating Prefect reviews this phase of debate on the social and economic issues and suggests the specifications for the infrastructure to the Minister.
The design studies are monitored with reference to these specifications, after government approval. The specifications are attached to the dossier submitted to the Public Inquiry

---

Figure 10.2 French approach for ensuring transparency in megaproject decision making. The approach has been institutionalised by means of a government circular
Source: Jean-Michel Fourniau, 'Making the Decision More Transparent: Recent Changes in the Treatment of Major Transport Infrastructure Projects in France', unpublished paper, INRETS-DEST, Arcueil, France, undated, p. 11.

and democratic process stand to benefit from the use of traditional and extended peer review. Such review should be applied to feasibility studies as well as to decision-making and implementation processes. Substantively peer review should cover estimates of costs and revenues, viability, financing, regulatory regimes, safety and environmental issues, compensation, mediation, implementation strategies, and so on. Independent peer reviewers should write their own reports and press releases. In countries with a good and trusted Auditor-General, ongoing review of economic and financial matters may be placed there. Besides peer review, the organisation of scientific conferences about projects may help enhance the knowledge base for such projects.

Danish Ministry of Transport officials have talked about a 'democracy deficit' in the decision-making processes regarding the Øresund and Great Belt projects, and about the need to reduce this deficit by introducing more democracy in decisions about future projects, for instance the Fehmarn Belt link. We agree with this analysis and see the enforcement of the test of public scrutiny as the main means for eliminating the democracy deficit.

Performance specifications

The use of performance specifications implies a goal-driven approach to feasibility studies and decision making, instead of the conventional technical solution-driven one (see Figure 10.3). The use of a performance specification approach means that, to the extent possible, all requirements with respect to a possible project are to be decided before considering various technical alternatives for concrete solutions and before appraising the proposed project.[12]

In engineering, the performance specification approach has in recent years become more common in the development of various types of facility, thereby replacing in part the traditional approach, which is based on detailed technical specifications.[13] In our judgement, this way of thinking should be taken further by applying it not only to technical aspects of projects but also, for example, when considering their external effects.

In principle, performance specifications would derive from policy objectives and public interest requirements to be met by the project, for instance regarding economic performance, environmental sustainability and safety performance. More specifically, a transport project's performance may relate to such things as road and road link safety, safety to passengers (for example requirements with respects to rescue operations in tunnels, and so on), maritime safety (navigational aspects),

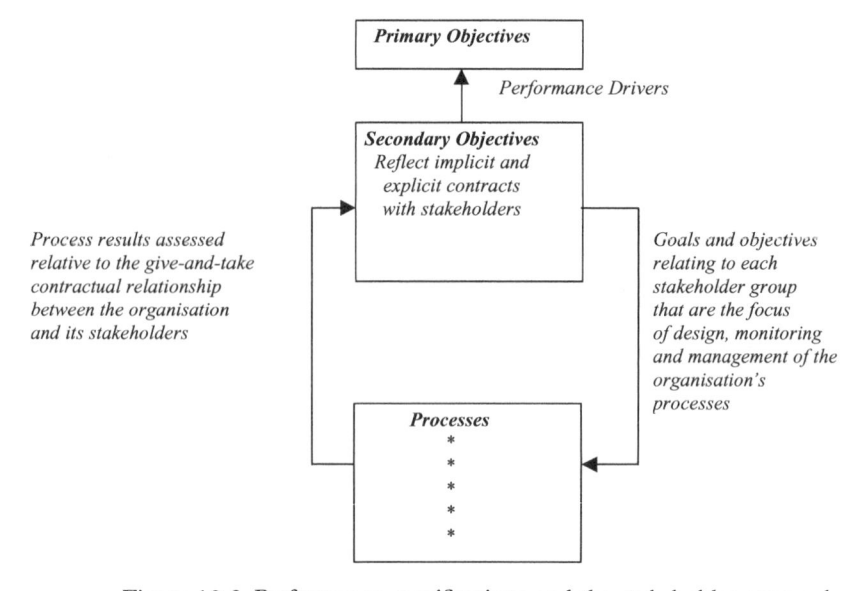

Figure 10.3 Performance specifications and the stakeholder approach in combination
Source: Anthony A. Atkinson, John H. Waterhouse and Robert B. Wells, 'A Stakeholder Approach to Strategic Performance Measurement', *Sloan Management Review*, vol. 38, no. 3, spring 1997, p. 31.

environmental impact (emissions, marine environment, energy consumption, and so on) and restrictions on land connections (for example identification of possible locations). Performance may also relate to such things as the capacity of a road or rail link, minimum and maximum speed of vehicles, and so on.

Performance specification thus covers more than just the approach to the development of a project from a technical point of view. Performance requirements will reflect national objectives in the transport and environmental sectors, among others. However, it is a requirement that specifications are formulated consistently and that they can be measured in an unequivocal way. The measurement requirement is necessary to enable the detailed design to be undertaken, and to allow for monitoring and auditing during possible construction and operation.[14]

One advantage of using a performance specification approach is that it forces people to focus on the ends rather than the means, namely on questions of what we want to achieve or want to avoid rather than technical questions. What level of safety should be required? Which are the environmental concerns, and how can they be addressed? In the

Øresund case, for example, a major issue turned out to be the effects of the flow of water through the sound, and it was eventually agreed that the fixed connection across Øresund must be constructed in such a way that there would be practically no impediment to the flow of water, so as to guarantee the quality of the marine environment in the Baltic Sea. This can be seen as a performance requirement to be met by a fixed connection across Øresund. The requirement should have been formulated up front in the decision-making process, as the problem was already known from the Great Belt project. Instead it was formulated and dealt with at a late stage, creating public havoc and destabilising decision making.

Another point to be made is that not only should major requirements be set before a decision is taken, but also the performance specification process should be initiated even before major investigations have been undertaken as part of feasibility studies. The advantage of the performance specification approach is thus that it allows for a constructive and reflexive dialogue with those who play an active role with respect to environmental, safety, economic and other issues. At the same time, the approach forces organisations and groups of people to play a constructive role in relation to how to meet the objectives they would like to see met, and undermines the credibility of criticism directed at major projects simply because they happen to be major projects.[15]

As argued in the previous chapter, but repeated here on account of its importance, performance specifications – when applied through a competitive tendering process – are also advantageous from another point of view, as they allow for innovative technical designs and operational practices to be introduced on the initiative of bidders, which may result in considerable cost savings in comparison with the conventional approach. In conventional contracting, the final design is normally prepared before calling for bids, leaving little scope for introducing new solutions and cost-saving devices later. When applying performance specifications, the design and how it is prepared are in principle entirely up to the contractor; the only requirement is that the design meets the performance specifications. Indeed, in a genuine performance-based contract the contractors will be responsible for undertaking not only design and construction, but also operations for a longer period of time – normally not less than ten years – which gives an incentive to contractors to introduce innovative procedures and techniques as they are effectively offered protection from sharing the information about these new techniques and procedures until the end of the contract period. This contracting approach therefore also allows for savings of costs in a life-cycle perspective.[16]

Regulatory regime

The regulatory regime is here seen as encompassing not only the economic rules regulating the construction and operations of a possible major project and other economic rules which have a significant bearing on the financial and economic performance of the project, but also the rules regulating the complementary investments that will be required in order to ensure a rational use of the project. For a coast-to-coast fixed link or an airport, for instance, these complementary investments include access links, and the regulations will cover issues such as whether these links will require public funding or will be financed by tolls from the main project.

One reason why this regime should be specified up front as far as possible is that it will make the government carefully review the issues under this heading, and identify all costs before any decisions are made. Another reason is that the choice of regulatory regime will significantly influence the risks of the project, and both costs and risks should be central to any feasibility study and appraisal.

Finally, if part of the financing for a possible project is to be mobilised from genuine risk capital this can only take place if the regulatory regime is set out, and risks which are of a political nature are identified, and, where relevant, as far as possible eliminated. The regulatory regime also involves such questions as the need for price regulation, that is whether the prices to be set by a possible future operator for use of a facility should be subject to some kind of control and whether competitive services – for example ferry boats in the case of a bridge or tunnel – should be allowed and to what extent.[17]

The need for price regulation will have to be reviewed carefully, as such controls often give rise to economic costs and may breed unexpected inefficiencies.[18] It should be recognised that existing competition legislation provides a powerful tool for combating dominant behaviour, and that such behaviour is also controlled by allowing for competing services. Economic regulatory requirements for the entity operating a given facility may consequently have to be very limited, for example restricted to requiring disclosure of financial information relevant to the setting of tariffs and the like.

For transport infrastructure, there is a particular importance in identifying all associated costs since, among other things, access rail and road links will often have to be built and will, possibly, require public money. Those who pay the money, namely the public, should know the full consequences in this regard. Again, Øresund and Great Belt can serve as examples. After Denmark and Sweden had agreed to build the Øresund link, it was realised that the envisaged access railway on the Swedish side would not be adequate, at least in a longer-term perspective. The idea of providing an alternative, and additional, new access link – the so-called City tunnel through Malmö – was therefore promoted and approved in 1997. The cost has been estimated as in excess of SEK 7 billion (1996 prices, including technical contingencies but excluding interest during construction), that is costs equivalent to about 30–40 per cent of the cost of the fixed link itself, and more than twice the cost of the originally envisaged access links which have actually been built. All this money will, in principle, come from public funds.[19]

Similarly, when the Great Belt rail link was about two-thirds through construction, the Danish State Railways claimed that capacity problems would arise on the access links after it opened, and that either capacity would have to be increased on these links or the result could be that one third fewer trains than expected would use the link. After the rail link opened in June 1997, capacity indeed became a problem. The result has been a decrease in reliability of train services throughout Denmark. Using the Danish State Railways' own definition of a delay (a departure or arrival that is more than five minutes behind the timetable), the number of delays for trains leaving Copenhagen towards Great Belt increased by 233 per cent – from 3.3 per cent of all trains to 11 per cent – in the six months after the opening of the rail link as compared with the six months before.[20] Long delays (larger than ten minutes) increased by 325 per cent. In Odense, a provincial capital close to the link and the most important bottleneck on the east–west rail route of which the link forms part, a full third of all departing trains bound for Great Belt were delayed after the opening of the link compared to 13 per cent before. Long delays in Odense increased by 162 per cent. With time some of the delays have been eliminated, but reliability has generally been lower after the Great Belt rail link opened compared with before. Hundreds of millions of kroner have been spent developing and debating solutions to the capacity and reliability problems, but no answer has been settled on as yet. Three things are certain, however. First, it will be highly costly to solve the problems; costs of up to DKK 10 billion are quoted, that is 40–50 per cent of the cost of the fixed link. Second, even if a decision was made now on how to solve the problems, the lead time for an engineering project of this size would be an estimated eight years. Third, during this period the reliability of train services in Denmark will continue to be in jeopardy.

As if these problems were not enough, when the Øresund link between Denmark and Sweden opened in July 2000, the problems with congestion and delays became even worse, with more trains on crowded tracks, especially in and around Copenhagen.[21] Again it will take several years and large sums of funds before these problems can be solved. As a result the public's trust in Danish railway services has been damaged. The irony is that the Great Belt and Øresund links were promoted as historical projects that would create a renaissance for rail in Denmark by strengthening the competitive position of rail services *vis-à-vis* roads. Clearly, an informed decision about the Great Belt and Øresund links would have had to take the capacity problems on access links – and the costs and regulatory regime for solving them – into account.

Risk capital

In principle, the most important issue from an accountability point of view is the actual decision on whether to undertake the investment in a megaproject or not. Given that the size of the investment is in the multibillion-dollar range and that the uncertainties involved are substantial, it seems self-evident that it must be possible to hold accountable those who take the decision. Experience from Great Belt, Øresund and other projects shows that government in itself is not sufficiently effective when it comes to enforcing accountability with respect to specific issues such as decisions on major infrastructure investments (see Chapters 2–4). A more effective way of achieving this is, in our judgement, to let the decision to go ahead with a project – given that the project satisfies agreed public interest objectives – be conditioned by the willingness of private financiers to participate in the project without a sovereign guarantee. This means that at least part of the capital, which will have to be mobilised for a given project, should be genuine risk capital. In other words, only if this risk capital can be mobilised will the project be built. By requiring that a substantial commitment in the form of risk capital is made, the ordinary citizen will not be required to carry any, or only limited, risks. The common practice, followed at Øresund and Great Belt, of transferring the costs of risk to those who are in the weakest position to protect themselves is thereby, if not eliminated, at least significantly reduced.[22]

However, even if income distributional effects are disregarded, and focusing only on the pure economic merit of an investment such as an infrastructure project, we would argue for the merits of having risk capital involved. Investments in major infrastructure projects are resource allocation decisions with impacts many years into the future. To mobilise the finance required, ultimately money set aside by ordinary people in the form of savings for the future will have to be drawn upon. The decision on whether to build or not will have to be taken subject to a high degree of uncertainty, and as elaborated upon in Chapter 7, there is no objective way of measuring the cost of that uncertainty. Under these circumstances, it is important that a decision to commit the savings of ordinary people is placed in the hands of people who can be held accountable. By making the decision conditional on private financiers' willingness to invest in a project, and by letting them bear the consequences of a wrong decision, there will be a better guarantee that a project will indeed only be implemented if there is a demand for it.

The involvement of risk capital will also have another beneficial effect if it is decided to go ahead with a project. It will ensure a higher degree of involvement by the lenders during the final design, construction and operation of the project, and more effective monitoring. As a consequence, better cost control can be expected and also better controls against construction delays. As demonstrated in a study by Eichengreen, the use of government guarantees when constructing railways in the last century was a mixed blessing, as such guarantees removed the incentive for investors to monitor management performance. The evidence from that period suggests that costs of construction works undertaken without guarantees were lower than for works, which had such support.[23]

An argument against the use of private risk capital that we sometimes encounter when discussing these issues with government officials and project promoters is that not enough private capital is available for infrastructure building. Thus many projects would not be realised if they were to rely on private funds, according to this argument. The argument is false, nevertheless, as any economist must concede, because both in principle and in practice there is abundant private capital available for profitable projects. If such capital is not forthcoming for a project this is because the market does not believe in it, and this is the whole point of involving private risk capital. This, of course, may in some cases be seen as a problem by contractors and consultants who make a living from developing and building projects, or by such politicians and administrators who would like to see their tenure capped by the monumental effect of building a megaproject. But we firmly hold that this is the way things should be in order to avoid the risk of taxpayers being held captive to the rent-seeking behaviour of special interest groups, be they after profit or fame.

The involvement of private capital further raises the question of the appropriate organisational structure for constructing and operating a given piece of infrastructure. So far, the model chosen in Denmark for the major fixed links has been that of a state-owned enterprise (SOE) established in terms of the Companies Act. That approach will also be possible when a degree of risk capital is involved; there is, in principle, no need for a government to guarantee the debt incurred by enterprises that it owns.[24] However, if risk capital is a condition for building a given infrastructure project, a higher financial return will be required for this project than for projects with sovereign guarantees such as the Øresund link, which was justified on the basis of a real rate of return of 5 per cent. In comparison, the Great Belt link was justified on the basis of a real rate of return of 10 to 15 per cent, depending on method and time of calculation.[25] As illustrated in Chapter 7, a proper risk evaluation would involve a required rate of return of about 9 to 10 per cent in real terms (disregarding taxation effects). It is uncertain whether it would have been possible to justify the Øresund link on the basis of a real rate of return of 9 to 10 per cent, and therefore the recommendation on risk capital may be controversial to the promoters of this project. On the other hand, it may be maintained that many of the substantial problems that have been associated with the Øresund project are related to inadequate appraisal, including an improper treatment of risk.

> The requirement for risk capital does not mean that government gives up control over any given project. On the contrary, it is under this condition that government can effectively play the role it should be playing, as the ordinary citizens' guarantor that environmental, safety and economic concerns are properly addressed, and that rent-seeking behaviour by special-interest lobby groups does not capture projects.

The approach to appraisal of fixed links that has evolved in Denmark in recent years has come to place increasing emphasis on financial viability, apparently at the expense of economic (social) viability. The emerging approach seems implicitly to argue that so-called Pareto improvements (not just potential but actual Pareto improvements) should be aimed for, namely situations that will leave no one, or very few, worse off – and some better off – as a consequence of the link. The zero solution for environmental impacts on the Baltic Sea can be interpreted in this way, as can the requirement that projects must be financed by those, and only those, who benefit directly from them. A logical extension of this approach is that in terms of risk, too, no one should be left worse off because of a project, that is the risk costs should also be borne by those who make use of the project. While actual Pareto improvements, where no one loses, are notoriously hard or impossible to achieve for large projects, it seems clear that the emerging approach to appraisal gets nearer to actual Pareto improvements than the conventional approach. The latter focused on potential Pareto improvements with no concern for the distribution of effects so long as winners would, in principle, be able to compensate losers, whether or not actual compensation took place.

It should be emphasised that to meet the requirement for effective accountability, it is not necessary that all financing for a given project is mobilised on the private market and without a sovereign guarantee. It would be possible to guarantee part of the financing required in order to lower somewhat the direct financing costs. In our judgement one third or more of the total financing requirements should be financed by private risk capital in order to ensure the acid test of viability and thereby protect the public interest.

Finally, it should also be re-emphasised that the requirement for risk capital does not mean that government gives up control over any given project. On the contrary, it is under this condition that government can effectively play the role it should be playing, namely as protector and promoter of the public interest, or, more specifically, as the ordinary citizen's guarantor that environmental, safety and economic concerns are properly addressed, and that rent-seeking behaviour by special-interest lobby groups does not capture projects.

Conclusions

In this chapter, we suggested a rearrangement of public and private responsibilities in megaproject development and appraisal, and we set out four basic instruments of accountability:

(1) *Transparency*. The test of public scrutiny is the main means for enforcing accountability in the public sector. All documents and other information should be available to the public. Stakeholder and civil society groups should be invited to participate from an early stage in feasibility studies and decision making. Government should take an active role in identifying such groups. Public hearings should be considered and used from an early stage. Independent peer reviews should be carried out for all important aspects of a project. The Auditor-General should be considered for financial and economic review of projects. Scientific conferences should be organised. Professional expertise should be used to plan and implement transparency and participation, not for PR purposes but for effective two-way communication, just as professionals are hired to secure the quality of most other tasks in major project development;

(2) *Performance specifications*. As far as possible, all requirements with respect to a project should be decided up front, before considering various technical alternatives and before appraising the project. Performance specifications would derive from policy objectives and

public interest requirements to be met by the project, for instance regarding economic performance, environmental sustainability and safety performance. The purpose is a goal-driven appraisal and decision-making process, instead of one dominated by discussions of technical alternatives. Performance specifications should be formulated in ways sufficiently concrete to allow for monitoring and auditing;

(3) *Explicit formulation of regulatory regime* (namely the economic rules regulating the construction and operations of a project, other economic rules with significant bearing on the financial and economic performance of the project, and the rules regulating the complementary investments which will be required in order to ensure a rational use of the project). This regime should be specified up front as far as possible in order to force government to review carefully the issues under this heading, and identify all costs before any decisions are made. Another reason is that the choice of regulatory regime will influence the risks of the project, and both costs and risks are central to feasibility study and appraisal. Finally, if part of the financing for a possible link is to be mobilised from genuine risk capital this can only take place if the regulatory regime is set out, and risks which are of a political nature are identified, and, where relevant, as far as possible eliminated;

(4) *Risk capital*. The decision to go ahead with a project should, where at all possible, be made contingent on the willingness of private financiers to participate without a sovereign guarantee for at least one third of the total capital needs. This would result in more realistic risk assessment, a possible reduction of risk and a shift in risk from ordinary citizens to groups better able to protect themselves against risk. The pressure on performance would be higher as lenders and possible shareholders and stock market analysts would monitor the project. The participation of risk capital does not mean that government gives up or reduces control of the project. On the contrary, it means that government can more effectively play the role it should be playing, namely as the ordinary citizen's guarantor for ensuring concerns about safety, environment, economics and distribution of risk are met.

11 Accountable megaproject decision making

With the necessary instruments for enforcing accountability elaborated in Chapter 10, we see two main alternatives for developing and reaching accountable decisions on whether or not to go ahead with a mega infrastructure project. The first alternative is based on the assumption that a project, if built, would be regulated by a concession, that is this alternative is based on the build-operate-transfer (BOT) approach, or a similar set-up. In the second alternative, a possible project, if built, would be constructed and operated by a state-owned enterprise (SOE). We assume in both alternatives that at least one third of the required capital will not be secured with government guarantees. We furthermore assume that the same taxation regime would apply to both alternatives.

The two main alternatives are set out in Tables XI.i and XI.ii, and should be compared to the conventional approach as summarised in Table VIII.i in Chapter 8. The first part of the process is similar for the two alternatives but differs towards the end, once all the public interest issues have been settled through the identification of performance criteria and regulatory provisions, and once it has been decided whether project implementation should be publicly or privately led. The steps involved in the development process would not necessarily follow each other in a simple consecutive order. Several of the steps would, most likely, be carried out simultaneously and with possible interactions and iterations.

The concession approach

The basic assumption of the concession approach is that, if it should be decided to establish a specific piece of infrastructure, private companies would be required to bid for a concession to build and operate it for a given period of time, and a contract would be negotiated with a selected consortium covering financing, detailed design, construction and operations.

Table XI.i *Alternative 1: the concession approach to project development*

Steps	Actions	Responsibility
1.	Undertake policy study; publish policy document	Government
2.	Prepare terms of reference; and recruit consultants to draft performance specifications	Government
3.	Prepare draft performance specifications based on government policy objectives, laws and regulations	Consultants
4.	Prepare terms of reference; recruit consultants to prepare feasibility study	Government
5.	Prepare terms of reference; recruit consultants to prepare plan for public involvement (public hearings, stakeholder group involvement, peer review, etc.)	Government
6.	Prepare pre-feasibility study; if study indicates an unfeasible project, the process may stop here	Consultants
7.	Prepare Consultation Document 1, to be used for wide consultations with public and stakeholders	Government
8.	Consultation with public, stakeholders and regulatory bodies	Government
9.	Prepare terms of reference; recruit consultants to: propose regulatory regime; do further analysis of additional, associated costs; prepare risk management plan; and make proposals for operation, etc.	Government
10.	Prepare Consultation Document 2 for wide consultation with public and stakeholders	Government
11.	Prepare Final Performance Specification Document	Government
12.	Prepare Decision Document to identify: • performance specifications • financing conditions for operation • risk management • mode of operation • tender procedures, if relevant • regulatory regime • cost estimates and financing conditions for additional associated costs	Government
13.	Develop necessary legislation and make decision in Parliament to stop or go ahead with project	Government/ Parliament
14.	If project is ratified, undertake pre-qualification of bidders	Government with assistance of consultants
15.	Prepare shortlist and ask for bids	Government with assistance of consultants
16.	Evaluate bids, including acceptance from performance point of view; if no bids received, or bids fail to meet performance specifications and bidders not willing to modify their bids accordingly, the process stops here	Government, including relevant regulatory bodies
17.	Select concessionaire, negotiate and sign preliminary agreement	Government with consultants

Table XI.i (*cont.*)

Steps	Actions	Responsibility
18.	Prepare and circulate Information Document; publication subject to review by Auditor-General; at this point selected concessionaire can initiate final designs to obtain: (i) final permits from regulatory authorities (ii) bids from contractors	Government and concessionaire
19.	Submit negotiated agreement for approval and signature by relevant authorities and concessionaire	Concessionaire and government
20.	Prepare detailed design and obtain final clearance from environmental and safety authorities; if clearance not obtained the project may be terminated at this point	Concessionaire and government
21.	Implement agreement	Concessionaire
22.	Monitor and audit agreement	Government

Table XI.ii *Alternative 2: the state-owned enterprise approach to project development*

Steps	Actions	Responsibility
1–13.	Same as in Table XI.i	
14.	If project is ratified, establish state-owned enterprise(SOE)	Government
15.	Identify financial performance requirements to be met by SOE; negotiate preliminary agreement for these requirements	Government
16.	Require SOE to negotiate preliminary agreement with potential financiers; if agreement not reached, project may be terminated at this point	Government
17.	Prepare and circulate Information Document; publication subject to review by Auditor-General	Government
18.	Submit agreement between government and SOE for ratification and signature by relevant authorities and SOE	Government
19.	Implement agreement	SOE
20.	Monitor and audit agreement	Government

Whether or not the circumstances of a given piece of infrastructure, say an airport, a bridge or a tunnel, would provide for a build-operate-transfer approach as a serious contender for providing the services in question must be the subject of concrete, contextual analysis. However, given the fact that the provision of transport services can often be defined and organised as a fairly straightforward commercial activity, and granted that there are often alternative routes and competition for traffic between two geographical points, a build-operate-transfer approach is likely to show itself to be a relevant alternative in a number of contexts, for instance when building bridges and tunnels across waters where competing ferry lines will be in operation after a fixed connection has been established.

Nevertheless, we wish to sound two warnings regarding the build-operate-transfer approach. The use of this approach requires full commitment from the government or governments involved. The basic idea of the approach is to carefully think through essentially all the issues related to a project and to work out clear strategies for how to address them before any final commitment is made. If one or more governments involved in a project would have their doubts as to the relevance and appropriateness of the approach, or if there were strong political opposition, then it would probably not be prudent to embark on such an approach. The other warning concerns the need for ensuring that the interests behind a possible concession company are of a long-term nature, namely that they are not mainly related to construction, and that these interests are firmly entrenched in the ownership and management structure from the very beginning. Experience shows this to be crucial, as demonstrated in the previous chapters.

The state-owned enterprise approach

We mentioned previously that the model chosen for the Great Belt and Øresund projects has been that of a state-owned enterprise (SOE) established in terms of the Companies Act. We also mentioned that this approach does not eliminate the possibility that some element of risk capital is involved; in principle there is no need for a government to guarantee the debt incurred by enterprises that it owns. An advantage of this approach, in comparison with the concession approach, is that in principle it offers better opportunities for ascertaining that the operator has an appropriate management structure to safeguard the interests of the project company in the long term.

Criticism can be, and has been, advanced against state-owned enterprises like the ones established for implementing and operating the fixed links across Great Belt and Øresund, for example as concerns the transparency of the management of these companies and how risk has been handled. However, given the view that a fixed connection basically produces an ordinary service, it is appropriate to implement and operate a link through a company. The more important aspect is to establish clear performance standards for that company and to ensure that the company is appropriately staffed from the start, and that the members of the board of directors are adequately competent and are conversant with the implications of performance management and the notion of accountability. By ensuring that the four means of accountability identified in Chapter 10 are fully satisfied, including transparency, there should in other words be no fundamental problem with making use of a state-owned enterprise.

> In principle there is no need for a government to guarantee the debt incurred by enterprises that it owns.

As said, a state-owned enterprise should be required to mobilise part of the required capital without any guarantees. There are two ways of achieving this. Either a minimum of one third of the cost of the project is borrowed on the capital market without any guarantees, or part of the private capital is mobilised in the form of equity, leaving a correspondingly smaller balance to be borrowed without guarantees. In the latter alternative, the state-owned enterprise would therefore be partly privately owned, but the government(s) involved would be the majority owner. The benefits of mobilising private capital in the form of equity are underscored in a review of experience with public joint-stock companies in Denmark, published by the Danish Ministry of Finance. The review concludes that selling shares from a public joint-stock company to private investors entails a number of advantages in terms of incentives for better performance:[1]

(1) Public sale of shares through the stock exchange would make the company interesting to, among others, stock market analysts, thus creating in the market continuous critical review of, and pressure on, company performance;
(2) Representatives of private owners may gain seats on the company's board of directors. As members of the board, they would be better able than politically appointed members to oppose politically determined positions that could go against company interests;
(3) Going public on the stock exchange would enforce stricter rules regarding information to the board of directors of the stock exchange;
(4) The sale of shares could facilitate strategic alliances with relevant partners to improve and secure performance;
(5) The state is traditionally seen as a 'patient owner'. Private ownership would put more pressure on management to perform.

Selling shares from a public joint-stock company to private investors is therefore a means for promoting accountability.

> Documents should be made available to the public as they are produced and should be subject to independent and public peer review on major issues.

The project development process

Both of the alternatives above involve the preparation of a number of documents which should be made available to the public as they are produced in order to ensure transparency, and which should be subject to independent and public peer review on major issues, as has been the case for certain environmental issues on the Great Belt and Øresund projects. These documents are:

- *Basic Policy Document*: This is the initial and initiating document in which the government identifies all the issues associated with the proposed project and sets out a strategy for how to deal with them, in particular the process for formulating the performance specifications, and involvement of the public;
- *Draft Performance Specifications Report*: This is a document to be prepared by consultants, commissioned by government, which would identify all issues related to the formulation of performance specifications, and which would come up with a first proposal for how to set the performance standards. The document, which should be published, would, *inter alia*, cover:
  (a) the services and infrastructure to be considered during appraisal;
  (b) technical specifications from a performance point of view;
  (c) location of additional, associated infrastructure, for example access links;
  (d) financial and economic requirements;
  (e) environmental requirements;
  (f) safety requirements;
  (g) other requirements.
- *Pre-Feasibility Study*: This is a document to be prepared by consultants, commissioned by government, and which would be close to a full-scale feasibility study, excluding considerations on detailed design of alternatives. The assessment of economic viability would include all related investments and operational costs, for example with respect to additional, associated infrastructure, that would have to be incurred on

account of the core project. The document, which should be published, would, *inter alia*, cover:
(a) preliminary identification of a few basic alternatives;
(b) preliminary design of these alternatives, including cost estimates;
(c) market study (travel demand forecasts in the case of transport infrastructure);
(d) preliminary identification of supplementary investments required, including cost estimates;
(e) preliminary economic and financial viability evaluation;
(f) preliminary risk analysis using the MLD-principle (Most Likely Development), break-even and worst-case scenarios (see Chapter 7).

The aim of the study would be to assess the financial and economic viability of alternative solutions to the problem at hand. If, at this stage, it seems highly unlikely that such solutions would be viable, the process would then stop. If the conclusion of the study were that there could be merit to the project, preparation would proceed to the next phase.

> If it turns out that performance specifications have been set in ways that cannot be achieved, or if the project is politically infeasible in other ways, the process would be stopped here.

- *Consultation Document 1 (performance specifications)*: This document would justify, to begin with, why government considers that preparation should proceed, and should be based on the findings of the pre-feasibility study. It would then comprise a list of performance specifications that are believed to be relevant to the project. The document would be circulated and would then, together with Consultation Document 2 (see below), provide the basis for consultation, in the form of a remit process, feedback from stakeholder and civil society groups and public hearings. Input from this process would be used to revise performance specifications, where relevant. The Consultation Document would also be used as a first means of communication with the regulatory bodies responsible for granting required permits. If, after this process, it turns out that performance specifications have been set in ways that cannot be achieved, or if the project is politically unfeasible in other ways, the process would be stopped here.
- *Consultation Document 2 (risk management, operations, regulations, subsidiary financing)*: This document would make proposals for risk management, the future operations of a possible project, the level of

non-guaranteed private financing required, the type of economic regulatory regime to be applied, estimates of additional costs for supplementary investments, and proposals for the financing of these costs with a justification. The document would also be used in order to allow the public to identify risks. Non-experts, unencumbered by professional paradigmatic blinders, often see risks missed by experts. For remit procedure, and so on, see Consultation Document 1 above.

- *Final Performance Specification Document*: This is the document that would be submitted to the regulatory authorities in order to obtain preliminary clearance, and, where possible, preliminary permits to build the proposed infrastructure, if it is decided to go ahead with the project.
- *Decision Document*: This document would contain recommendations on all major policy issues (performance specifications, risk management, regulations, financing of subsidiary costs, operations, and possible tender procedures), and would thus identify all the essential conditions that a future operator would have to comply with. This document would also include, where relevant, draft legislation. It would be used for making decisions by Cabinet and Parliament on whether to go ahead or not with the project. If major changes in the project have taken place between Consultation Documents 1 and 2 and the Decision Document, the latter may be the object of another round of remit procedure, public hearings, and so on. Information produced by this process, and by earlier hearings would then be submitted to Cabinet and Parliament together with the Decision Document.
- *Information Document*: This would be the final major document to be prepared by the government, and would be issued either after negotiations have been concluded with a selected concessionaire, or when the state-owned enterprise has concluded its negotiations for mobilising finance. This is therefore the time when the selected operator can initiate preparation of final designs in order to obtain:

(a) final permits from regulatory authorities; and
(b) bids from contractors.

The purpose of the Information Document is to provide the public with full information as to how all the conditions set out in the Decision Document will be complied with. In view of the fact that the negotiated contracts cannot be revealed (as they are commercial documents), it would be appropriate for the government to allow the Auditor-General to audit the Information Document before it is released in order to ensure accountability.

Conventional and alternative approaches compared

Two positive tenets stand out for the conventional approach to major infrastructure project development, as evolved in Scandinavia in recent years: first, the basic notion that projects should not have negative impacts on the environment or on any (or very few) groups in society; second, that projects may be implemented in terms of the Companies Act, given the fact that projects often provide a fairly straightforward service that can be considered commercial. The company approach provides a framework for independent accounts to verify that financing requirements – self-financing, in the cases of Great Belt and Øresund – and other performance requirements are being met. It also establishes an operational format that in principle allows for efficient decision making by the organisation responsible for implementation and operations, but also for effective monitoring and auditing of this organisation by external parties.

The main argument for an alternative approach to the conventional model is that in the conventional approach the philosophy of not harming any group has not been carried to its logical conclusion, and that the definition and division of public *versus* company responsibilities is incomplete. The conventional approach gives rise to the following issues:

(1) The need to establish clear and unequivocal performance criteria. If the idea is that the decision should not have any marked negative impact, then this must be translated into explicitly stated performance criteria that must be attained by the project. These criteria should be formulated in such a way that they can be monitored and their attainment verified. Performance criteria should also be established well before any decision whether to begin construction or not is actually taken;

(2) The need to fully recognise the economic risks associated with megaprojects. Our review of a large number of projects shows that such risks are of considerable magnitude (see Chapters 2–5), are being ignored all too often and that a major reason for this appears to be that risks are being borne by the public at large. There seems to be the notion that just because the government is involved and the investment is ultimately underwritten by the taxpayers of a country, then there is no reason to be overly concerned with the risks, or even to recognise them publicly. This is a flawed way of reasoning; it also has major income distributional consequences. In line with the basic criterion that major projects should not have any marked adverse effects on any group in society, the investors and users of such projects

must ultimately be responsible for carrying the costs of uncertainty, not the taxpayers;
(3) The need to ensure that the decision-making process is not being captured by lobby groups. Special-interest groups play an important role in shaping policy in modern democracies, but they cannot be held accountable for decisions taken by government. Too often, decisions on major projects have been heavily influenced by specific interests, and, we would argue, at the expense of the general public. We have identified a need to limit rent-seeking behaviour by special interest groups, and we see three actions as required to achieve this:
   (i) full transparency during the decision-making process, which basically means full public access to documentation prepared by the government;
   (ii) full involvement of different stakeholder and civil society groups during the preparation and formulation of performance specifications; and
   (iii) that the final decision on whether to build or not is made dependent on the provision of private risk capital, that is capital which is not secured by the government. We recommend that such capital should constitute a minimum of one third of total capital requirements for a possible project.

The implications of the recommendation to make use of a new approach to project development are first and foremost to make the role of government substantially more clear, and to make it more possible to hold government accountable for playing its role. Such accountability is an explicit constitutional requirement in most western democracies. Therefore, at the most basic level, our proposal simply aims at bringing a certain political and administrative practice – that of megaproject development – closer to what, according to standard constitutional canons, it is already supposed to be. The role of government is ideally to serve and protect the public interest. The weakness of the conventional approach is that it allows government to play more than one role, for example as project proponent and as auditor of safety, environment and financial issues. As a consequence conflicts of interest are generated within government, accountability is diluted and the decision-making process becomes open to capture by special interest groups.

Legislating project development

The two main alternative approaches elaborated upon here should be seen as examples of how to institute accountability. We recognise that there may be other approaches, which fulfil necessary requirements for accountability. In particular, there may be other ways of structuring reporting and the process of communicating with stakeholders, civil society and the general public.

In many countries, the legal foundation for building and operating a major infrastructure project is set out in a public works act, which in legal terms defines the rules and procedures to be followed in the project during and after construction. In view of the magnitude, complexity and potential consequences of most major projects, rules and procedures should be set out in this manner not just for possible construction and operations, but also for project development and appraisal. Therefore, we suggest that the approach taken to project development and appraisal – whatever this approach may be – is set out in a government directive, or similar regulation, to be followed by those responsible for development and appraisal. This, too, would help improve accountability. It would also encourage the use of a consistent approach to project development and appraisal, which today is often sorely lacking.

12 Beyond the megaprojects paradox

The point of departure for this book was a paradox. Recent years have witnessed a steep increase around the world in the magnitude, frequency and geographical spread of megaprojects, namely multibillion-dollar infrastructure projects such as airports, high-speed rail, urban rail, tunnels, bridges, ports, motorways, dams, power plants, water projects, oil and gas extraction projects, information and telecommunications technology systems, and so on. Never in the history of humankind have we built more, or more expensive, infrastructure projects. And never have such projects been more central to establishing what sociologist Zygmunt Bauman calls 'independence from space' and Microsoft chair Bill Gates 'frictionless capitalism'. Yet when actual *versus* predicted performance of megaprojects are compared, the picture is often dismal. We have documented in this book that:

- Cost overruns of 50 per cent to 100 per cent in real terms are common in megaprojects; overruns above 100 per cent are not uncommon;
- Demand forecasts that are wrong by 20 per cent to 70 per cent compared with actual developments are common;
- The extent and magnitude of actual environmental impacts of projects are often very different from forecast impacts. Post-auditing is neglected;
- The substantial regional, national and sometimes international development effects commonly claimed by project promoters typically do not materialise, or they are so diffuse that researchers cannot detect them;
- Actual project viability typically does not correspond with forecast viability, the latter often being brazenly over-optimistic.

> The aim is to decrease the risk of governments, taxpayers and private investors being led – or misled, as often turns out to be the case – repeatedly to commit billions of dollars to underperforming projects.

Causes of the megaprojects paradox

We have identified the main cause of the megaprojects paradox – namely the fact that more and bigger megaprojects are built despite their poor performance record – as one of risk-negligence and lack of accountability in the decision-making process. We have shown that project promoters, unsurprisingly, are happy to go ahead with highly risky projects as long as they themselves do not carry the risks involved and will not be held accountable for lack of performance. We have also shown that with the conventional approach to megaproject development, all too often promoters have actually been able to dodge risk and accountability. Finally, we have proposed measures and institutional developments to curb this problem. The aim is to decrease the risk of governments, taxpayers and private investors being led – or misled, as often turns out to be the case – to repeatedly commit billions of dollars to underperforming projects.

Clearly, nobody has an interest in risky and underperforming projects in and of themselves. However, contractors and other project promoters who stand to gain from the mere construction of projects, and who are often powerful movers in the early stages of project development, may have a self-serving interest in underestimating costs and overestimating demand, and in similarly underestimating environmental impacts and overestimating developmental effects. Construction periods for megaprojects are often decade-long, relegating the concerns of operations to a distant future. The period from project proposal to start of operations is even longer. The tactical under- and overestimation of effects in the initial stages of project development make projects look good in the cost–benefit analyses and environmental impact assessments that today invariably accompany megaproject proposals. This, in turn, increases the likelihood that projects will be built and thus benefit contractors and other construction-oriented promoters. Undoubtedly, this explains many decisions to build projects that would not have been given the go-ahead if more balanced appraisals had been available, if the concerns of operations had been given more weight and if the promoters had been held accountable for their claims about projects.

Such rent-seeking behaviour and the associated 'appraisal optimism' – to use a polite World-Bank term for biased cost–benefit analysis – are not in the interest of those whose money is put at risk, be they taxpayers or private investors. Nor are they in the interest of those concerned with environment, safety, democracy and the public interest. For people and groups who, like us, share such concerns, we propose the measures described below to increase accountability in megaproject decision making.

The measures are explained in depth and are detailed for actual use in the body of the book. In addition, the book's Appendix contains an example of the measures put to work for a specific project with which we have been involved as advisers, namely the proposed Baltic Sea link connecting Denmark and Germany across Fehmarn Belt, one of the largest cross-national infrastructure projects in the world.

> The tactical under- and overestimation of effects in the initial stages of project development make projects look good in cost–benefit analyses and environmental impact assessments.

Cures for the megaprojects paradox

Our basic, and first, proposal is that risk and accountability should be much more centrally placed in megaproject decision making than is currently the case. We see good decision making as a question not only of better and more rational information, but also of institutional arrangements that provide the checks and balances necessary to ensure accountability, here especially accountability toward the substantial risks that we have shown exist in megaproject development. Financial, environmental and safety risks cannot be eliminated from megaprojects, but they can be acknowledged and reduced through careful identification of and allocation to those best suited to carry them. To this end we recommend that laws or similar kinds of regulation should be passed stipulating that risk analysis and risk management must be carried out and must be institutionally anchored in megaproject decision making, with risks allocated to relevant actors and organisations. We also recommend and detail methods for risk analysis, including what we call Most-Likely-Development (MLD) analysis, break-even and worst-case scenarios. Finally, under this heading of accountability we recommend institutional arrangements for more effective risk management.

Our second proposal is a rearrangement of public and private responsibilities in megaproject development. In what we call the conventional approach to megaproject decision making, government plays a host of roles, some of them conflicting. We ask the question, can a government act effectively both as promoter of megaprojects, and as the guardian of public-interest issues such as protection of the environment, ensuring safety and shielding the taxpayer against unnecessary financial risks? We answer this question in the negative. There is a conflict of interest for the government here, and as a result accountability suffers. We then redraw the borderlines of public and private involvement, shifting risk from the public to the private sector and establishing a substantially clearer role for government by means of the arm's-length principle and by shifting government involvement from project promotion to formulation and auditing of public-interest objectives to be met by megaprojects.

> There is little evidence that efficiency and democracy are trade-offs for megaproject decision making. Quite the opposite.

Thirdly, we propose four basic instruments of accountability to be employed in megaproject decision making:
(i) Transparency;
(ii) Performance specifications;
(iii) Explicit formulation of regulatory regime; and
(iv) The involvement of risk capital.

Transparency

Transparency is the main means of enforcing accountability in the public sector. Here we recommend a higher degree of publicness and of public participation, including stakeholder involvement, than is common for megaprojects. Megaproject development stands to benefit from more involvement by civil society. The conventional argument against public participation is that it slows down decision making and results in suboptimal outcomes. We read the evidence quite differently. Megaprojects that have tried to get by without publicness and participation have often run into such heavy opposition that the decision-making processes were destabilised and second-best solutions to both procedure and outcome forced upon actors and projects. There is little evidence that efficiency and democracy are trade-offs for megaproject decision making. Quite the opposite.

Performance specifications

The use of performance specifications implies a goal-driven approach to megaproject decision making, instead of the conventional technical solution-driven one. The use of a performance specification approach means that, as far as possible, all requirements with respect to a possible project are to be decided before considering various technical alternatives for the proposed project and before appraising it. In principle, performance specifications would derive from national and other policy objectives and public-interest requirements to be met by the project, for instance regarding economic performance, environmental sustainability and safety performance. One advantage of using a performance-specification approach is that it forces people to focus on the ends rather than the means. It allows for a constructive and reflexive dialogue with those who play an active role with respect to environmental, safety, economic and other issues. At the same time, the approach forces organisations and groups of people to play a constructive role in determining how to meet the objectives they would like to see met, and undermines the credibility of criticism directed at megaprojects simply because they happen to be megaprojects.

Regulatory regime

The regulatory regime is the set of economic rules regulating the construction and operations of a specific megaproject, other economic rules that have a significant bearing on the financial and economic performance of the project, and the rules regulating the complementary investments that will be required in order to ensure a rational use of the project, for instance regarding access links in the case of airports or bridges, and whether such links will require public funding or will be financed by tolls. One reason why the regulatory regime should be specified up front as far as possible is that it will make government carefully review the issues under this heading, and identify all costs before any decisions are made. Another reason is that the choice of regulatory regime will influence the risks of the project, and both cost and risk should be central to any feasibility study and appraisal. Finally, if part of the financing for a possible project is to be mobilised from risk capital, as we propose it should be, this can only take place if the regulatory regime is set out, and risks which are of a political nature are identified, and, where relevant, as far as possible eliminated.

Risk capital

In principle, the most important issue from an accountability point of view is the actual decision on whether to undertake the investment in a megaproject or not. We have documented that government in itself is not sufficiently effective when it comes to enforcing accountability with respect to specific issues such as decisions on mega infrastructure investments. We propose that a more effective way of achieving accountability is to let the decision to go ahead with a project – given the project satisfies agreed public-interest objectives – be conditioned by the willingness of private financiers to participate in the project without a sovereign guarantee. This means that at least part of the capital that will have to be mobilised for a given project should be genuine risk capital. In other words, only if this risk capital can be mobilised will the infrastructure be built. By requiring that a substantial commitment in the form of risk capital is made, the ordinary citizen will be required to carry no, or only limited, risks. The common practice of transferring the costs of uncertainty to those who are in the weakest position to protect themselves is thereby, if not eliminated, at least significantly reduced. By making the decision conditional on private financiers' willingness to invest in a project, and by letting them bear the consequences of a wrong decision, there will be a better guarantee that a project will indeed only be built if there is a demand for it. The involvement of risk capital will also have another beneficial effect if it is decided to go ahead with a project. It will ensure a high degree of involvement by the lenders during the final design, construction and operation of a project, and more effective monitoring. As a consequence, better cost control can be expected and also better controls against construction delays.

Fourth and finally, we propose two alternative models for accountable megaproject decision making, one based on the state-owned enterprise (SOE) approach to project development, the other on the build-operate-transfer (BOT) or similar approach. Both are detailed to the level of, among other issues, stakeholder participation, environmental impact assessment and reporting procedures. We do not recommend one approach over the other, since the specific political and institutional context of a given megaproject will decide which approach is the more appropriate. We do suggest, however, that for any prospective project both approaches be given careful consideration before either is chosen. We also emphasise that independent of approach, it is crucial that clear performance standards are established up front for the project company, be it public or private, and that the company is appropriately staffed from the start. Construction interests must be balanced by operations interests among the members of the board of directors and within the staff from the very beginning, and both staff and directors must be conversant with the implications of performance management and the notions of risk and accountability. In short, if the four instruments of accountability outlined above are satisfied, either of the two approaches to megaproject development will do the job, and do it well.

Making risk society less risky

Megaprojects in transport, telecommunications and energy play a key role in nothing less than the creation of what many see as a coming new world order where people, goods, energy, information and money move about with unprecedented ease. Yet we have shown in this book that the track record of megaprojects is one of governments, taxpayers and private investors repeatedly being led to commit billions of dollars to highly risky, underperforming projects. We have demonstrated that the problem with megaprojects is mainly one of risk-negligence and lack of accountability induced by project promoters whose main ambition is to build projects for private gain, economic or political, not to operate projects for public benefit. And we have argued that in order to remedy the current state of affairs we must find ways of institutionally embedding risk and accountability in the decision-making process for megaprojects. Finally we have shown how this can be done.

We mentioned in the first chapter that theorists of risk society and of democracy have recently begun to call for the development of practical policy and planning measures that deal with risk in real-life public decision making. And we said at the outset that we offer this book as an attempt at fleshing out, for a specific domain of increasing social, economic and political importance, the type of planning and decision making called for by the theorists. Some of the measures we have proposed above are today slowly gaining a foothold in a few countries and megaprojects. We hope our book will help this foothold get stronger and will help the measures gain acceptance in more places and projects, thus making democracy more accountable and risk society less risky.

Appendix. Risk and accountability at work: a case study

Some time ago we were asked by the Danish Transport Council to undertake a review of plans to study the viability of fixed connections between Denmark and Germany across the Baltic Sea at Fehmarn Belt. The Danish Transport Council was set up by the Danish Parliament to improve the basis for Danish transport policy by funding research and by carrying out independent studies of current transport issues with policy recommendations to government, Parliament and the general public.[1]

Fehmarn Belt is the strait between Denmark and Germany located in the western part of the Baltic Sea between the islands of Lolland and Fehmarn. The shortest distance across the Belt is the 18.6 km (approximately 11 miles) between Rødbyhavn in Denmark and Puttgarden in Germany. A fixed connection across the Baltic Sea at this location would be one of the largest and most costly cross-national infrastructure projects in the world. Estimates suggest a cost in the range of DKK 20 billion to 35 billion (US$3.3 billion to 5.8 billion) for the coast-to-coast facility, and an additional DKK 15 billion to 30 billion (US$2.5 billion to 5 billion) for connecting access links. As part of an agreement with the Swedish government regarding the Øresund fixed link, the Danish government has committed to building a fixed Fehmarn Belt link, provided such a link is ecologically and economically feasible.[2] The position of the German government is less certain. The EU Commission has placed a fixed Fehmarn Belt link on a list of European transport infrastructure projects with high priority, which means the Commission can contribute 20 per cent to the investment. Because of certain problems with the development and appraisal processes for the Great Belt and Øresund projects (see Chapters 2–7), both comparable in size and purposes with the Fehmarn Belt link, the Danish Transport Council wanted to investigate how the decision-making process for this type of project could be improved upon.

In our review of the Fehmarn Belt plans, we found that the original proposals for feasibility studies made by the Danish and German Ministries of Transport largely coincided with what in this book has been termed the conventional approach to megaproject development (see Chapter 8).[3] We emphasised the need to strengthen accountability by making use of the four instruments elaborated upon in Chapter 10 above: more transparency, the use of performance specifications, better specification of regulatory regime and, finally, the use of private risk capital. We also illustrated how this could be done in the actual planning and appraisal process for Fehmarn Belt by developing alternative approaches similar to the concession and state-owned enterprise approaches described in Chapter 11.

The Danish Transport Council subsequently adopted our recommendations and made them the Council's own recommendations to the Danish government, Parliament and general public regarding how development and appraisal of proposals for a fixed connection across Fehmarn Belt should be carried out. In what follows we describe later developments in the Fehmarn Belt appraisal process and evaluate which of the Transport Council recommendations have found use in practice, and which not, with particular emphasis on the four instruments of accountability mentioned above. Our main focus is on the Danish situation because a similar review with similar recommendations was not carried out in Germany. The evaluation is a preliminary one, needless to say, since project development is still going on.

Main activities in the Fehmarn Belt development and appraisal process since the Transport Council adopted our recommendations in 1995 have been:

1995–99:	Feasibility studies.[4]
1999–2000:	Public hearing combined with brief independent peer review of feasibility studies.[5]
2000–02:	Call from the Danish and German governments for expression of interest from private companies regarding their possible involvement in providing financing for a fixed link.[6]

If, after the call for expression of interest, the governments decide to continue project development, remaining major tasks would be to finalise feasibility studies, conduct national and international hearings, make a political decision to build or not and, provided the decision is to go ahead, choose tender form and conduct the tender.

Increased private-sector involvement

The first and most basic recommendation of the Danish Transport Council was the rearrangement of public- and private-sector involvement in the project-development process and the introduction of private risk capital in the financing scheme for a possible project. Here the Fehmarn Belt project has come into agreement with the Council recommendations on several important points and differs substantially from the Great Belt and Øresund projects. During the 1980s and 90s, when the latter projects were decided and built, the politicians and officials involved were against private-sector participation in financing and risk taking for these projects. At first, this attitude prevailed for the Fehmarn Belt project as well. Thus Danish politicians and officials were initially against the Transport Council recommendations as regards this issue.

By the turn of the century this had changed, mainly because the Germans were unwilling to use public funds or a sovereign guarantee to finance the coast-to-coast link.[7] Private risk capital would have to be forthcoming. There is evidence that the Transport Council recommendations may have influenced the German position on this point. The recommendations for Fehmarn Belt were effectively communicated to the German Government and if the German demand for private risk capital for Fehmarn Belt was the result of a more general government policy seeking to promote private risk capital for transport infrastructure then one would expect demands for private capital to be forthcoming for many other projects than the Fehmarn Belt project. This is not the situation. On the contrary, for German rail projects, for instance, the share of public finance has even been increasing.

The media have, on several occasions, speculated that the emergence of the German demand for private financing of the Fehmarn Belt project was caused by German unification after 1989. During the 1990s, the new Germany had spent so many fiscal resources on unification, also in terms of transport infrastructure, that the government was forced to look for private capital to finance its activities, or so the argument goes. But if this were the case one would again expect that demand for private capital would be forthcoming for transport infrastructure projects as such, and, as said, this has not happened.

We conclude there is indication the Transport Council recommendations have influenced the German position on the use of private capital to finance the Fehmarn Belt project, although for reasons of complexity it would be difficult to provide final proof of such influence and its exact scope. As for other complex government decisions that unfold over several years and are affected by sets of different influences it is hard to say exactly what caused what and by how much.

At any rate, the Danes had originally hoped to use the financing model developed for the Great Belt and Øresund links, namely a state-owned enterprise backed by a sovereign guarantee. But so far Denmark has had to adapt to the German position. The call for expression of interest for private involvement in the Fehmarn Belt project mentioned above is the most tangible sign of this development. The call followed German Government regulations. The objective was to clarify private-sector interests in participating in the financing of the Fehmarn Belt link, that is to find out what it would take in terms of organisational and financial set-up to make the private sector interested. As part of the exercise, the private sector was given opportunity to document whether it would be able to carry out the activities involved in developing, constructing and operating a possible link as well as or better than the public sector.[8]

The future will show what the final arrangements will be for public and private involvement in the Fehmarn Belt project, should Denmark and Germany decide to build it. But involving the private sector as is currently happening, and ruling out a general sovereign guarantee, are major steps towards the Transport Council recommendations and away from the approach previously taken for this type of megaproject in this part of the world, including Great Belt and Øresund. As explained in the main body of the present book, the potential winners from this development are ordinary citizens as taxpayers because of decreased risk and increased accountability. The potential losers are officials with a vested interest in the state-owned enterprise model plus rent-seeking private developers.

Transparency and public participation

The Transport Council recommends that project development and appraisal for the Fehmarn Belt project be more open to public scrutiny and participation than was the case for the Great Belt and Øresund links. Important progress has been made on this point, especially when compared with Great Belt and Øresund at similar stages of development. More can be done, however, and the Council recommendations are far from met as yet.

So far the main activities regarding transparency have been the publication of two reports summarising the results from the feasibility studies, a brief independent peer review of these studies, a public conference and hearing presenting the results of the peer review, a report from the conference and, finally, a temporary Ministry of Transport website where anyone could comment on and discuss a possible Fehmarn Belt link.[9] What is lacking, when compared with the Transport Council recommendations, is stakeholder and interest-group involvement, more extensive use of public hearings and more extensive publication of technical documentation, for instance as regards traffic demand modelling and other elements of cost–benefit analysis and environmental impact assessment.

It was a problem for the 1999–2000 independent peer review that all necessary documentation was not available to the experts carrying out the review.

One reason for the slow progress in opening up the development and appraisal process to the public is, undoubtedly, simple inertia. The organisations and professions involved in transport policy and planning are less used to a participatory and open approach to planning and policy making – and more used to a rationalistic, expert-oriented one – than are organisations and professions in other fields, for instance city and regional planning. The Danish Ministry of Environment and Energy, responsible for city and regional planning, commented in 1996 that the planned procedure for appraising the Fehmarn Belt project was different from the procedure used in city and regional planning by not allowing for citizen participation in establishing the content of the appraisal and by not establishing a minimum timeframe for public hearings.[10]

Transparency is the main means for enforcing accountability in the public sector. When transparency is reduced, the losers are ordinary citizens whose possibilities for holding government accountable are diminished. The winners are those politicians and officials who prefer to carry out their activities without public scrutiny and interference. When judged against the Transport Council recommendations, such politicians and officials have until now had too much influence on project development and appraisal for Fehmarn Belt as regards transparency.

Performance specifications

The purpose of the Transport Council recommendations regarding performance specifications is to establish a goal-driven development and appraisal process. Here, so far, the Fehmarn Belt project is a case of opportunity lost. The appraisal process has been quite conventional, dominated from the outset by discussions of technical alternatives – bridge *versus* tunnel, four-lane road *versus* two-lane road, rail *versus* road, and so on – instead of by deliberation about policy objectives and public-interest requirements to be met by the project.

This state of affairs could be changing with the attempt at involving the private sector in financing and organising the project. As a point of departure for the call for private-sector interest discussed above, the Fehmarn Belt project, like other projects following this procedure, must be described regarding how and on what terms it will contribute to overall social development. What this means as regards performance specifications remains unclear, however. And after almost ten years of pre-feasibility and feasibility studies dominated by discussions of technical alternatives for how to cross Fehmarn Belt, the introduction of performance specifications at this stage stand in danger of becoming an instance of too little too late. Again the result is less accountability and higher risk.

Regulatory regime

The Transport Council recommends explicit formulation of the regulatory regime that would apply to the construction and operation of a possible fixed link across Fehmarn Belt. So far no such regime has been specified, apart from statements of intention by the Danish and German governments that access links on land may be financed by public funds, whereas such funds will not be available for the coast-to-coast connection. In addition, alternative regulatory regimes were described in outline in the expression of interest for private involvement, without preference given to one regulatory regime over the others.[11] Because the choice of regulatory regime affects risks and costs, proper feasibility studies and appraisals cannot be finalised until the regulatory regime has been decided. To be valid, the feasibility studies published in 2000 will have to be recalculated once this happens.

With the two governments' invitation to involve the private sector in project development and appraisal, progress has been made regarding the specification of regulatory regime. A major purpose of private involvement is precisely to clarify which types of regime are useful and possible. At the time of writing this book it was too early to predict the outcome of this process. But seen from the point of view of the Transport Council recommendations, things are moving in the right direction, though with substantial delay and with an inverted order between feasibility studies and the choice of regulatory regime.

Risk analysis

The Transport Council recommends a full financial risk analysis of a possible Fehmarn Belt link. Initially, such analysis was not part of the Ministry of Transport's plans for feasibility studies.[12] Based on the Council's recommendations, however, members of Parliament and the media put pressure on the Ministry in calling for risk analysis. Eventually the Ministry gave in.[13] The results of risk analysis were published in 1999 and recommendations for further analysis were presented in 2000 as part of the independent peer review and public hearing mentioned above.[14] Once the regulatory regime for a possible link has been decided, and costs and risks can therefore be assessed more accurately, then risk analysis must be redone to be valid.

But even in its current form, financial risk analysis of the Fehmarn Belt link is a major step ahead compared to the Great Belt, Øresund and similar projects. Such analysis is now on the agenda for large transport projects in Denmark, which was not the case before Fehmarn Belt and the Transport Council recommendations. Implausible as it may sound, financial risks were largely ignored in the planning stages for the multibillion-dollar investments at Great Belt and Øresund and until such risks caught up with the projects during construction and operations (see Chapters 2–4). For Fehmarn Belt, risks are acknowledged, and deliberation about risk has become part of the appraisal and decision-making processes. The winners are those whose money will be placed at risk, that is taxpayers and/or private investors. The losers are those promoters and developers whose main interest lies in the construction of projects and not in operations. They have an interest in ignoring risks, just as they have an interest in underestimating other costs and in overestimating benefits, because this increases the likelihood that projects are built.

Environment and socio-economic effects

For environmental impact assessment, the main difference between the Transport Council recommendations and the actual assessment of a possible Fehmarn Belt link lies in the fact that stakeholder groups and the general public have not been involved in the assessment process to the degree recommended by the Council. There are also important similarities, however. First, environmental impact assessment is one of the few areas in the feasibility studies where performance specifications are actually used as a basis for assessment, as recommended by the Transport Council. The most notable performance specification is the so-called Zero Solution, namely the requirement that the flow of water in and out of the Baltic Sea must be unchanged so as not to impact the fragile ecology of the Sea. Second, environmental impact auditing is expected to play a key role in environmental risk management for a possible Fehmarn Belt link, also as recommended by the Transport Council.

When the Transport Council first presented its recommendations in 1995, there was widespread concern that fixed links across Fehmarn Belt, Øresund and Great Belt might negatively impact on the marine environment in the Baltic Sea. Marine environmental risks were perceived as high (see Chapter 5). The links across Great Belt and Øresund were still being built and data were unavailable or inconclusive regarding real environmental impacts and risk. At the time of writing this book, such data had been produced by the environmental auditing programmes at Great Belt and Øresund. The data indicate that with due precaution – and state-of-the-art environmental risk management – it is possible to plan and implement major tunnels and bridges with consideration for the marine environment, even in a sensitive habitat such as the Baltic Sea.

Empirically based, *ex post* environmental auditing combined with environmental management has proved to be more important in safeguarding the environment than has hypothetical, *ex ante* environmental impact assessment – as maintained by the Transport Council and as argued in this book. The use of environmental audits is, quite simply, an all-important step toward risk management and accountability. This step was taken with the Great Belt and Øresund projects, and if the Fehmarn Belt link is built it will most likely follow the good example of its two predecessors in this respect.

For regional and economic growth effects, there is little difference between the Transport Council recommendations and the Fehmarn Belt feasibility studies. The Council concludes that socio-economic studies are not likely to prove particularly rewarding because study after study of other projects has already shown that the substantial regional, national and/or international development effects routinely postulated by promoters of projects typically do not materialise, the main reason being that in modern economies transport costs constitute a marginal part of the final price of most goods and services. Only highly localised effects should be studied, together with issues of possible compensation, according to the Council. The Danish Ministry of Transport seems to agree and little has been done in this area as part of feasibility studies. What has been done is mainly a rehash of earlier studies.[15]

Summary and conclusions

In sum, the degree to which the Transport Council recommendations for Fehmarn Belt have been met so far by actual developments in the project appraisal process is high for the use of private risk capital, medium for transparency and public participation, low for the use of performance specifications, low for specification of regulatory regime, high for the use of financial risk analysis, high for environmental impact assessment and post-auditing, and, finally, high again for socio-economic and regional development effects.

Although, as shown above, there is still scope for substantial improvement in all the mentioned areas, and especially for transparency, performance specifications and regulatory regime, the Fehmarn Belt appraisal process is a major step forward when compared to the situation just a few years ago when the Øresund and Great Belt projects were appraised.

Learning is clearly taking place. Judged on the basis of the Fehmarn Belt project there is reason for optimism regarding the possibility of changing development and appraisal processes for megaprojects in the direction of less risk for the ordinary citizen and more accountability for government, promoters and developers. Whether such optimism is ultimately warranted, only the future will show. The most important decisions for Fehmarn Belt still lie ahead, including the most basic one: whether to build or not.

Notes

I THE MEGAPROJECTS PARADOX

1. On the role of the European Union as a promoter of megaprojects, see John F. L. Ross, *Linking Europe: Transport Policies and Politics in the European Union* (Westport, CT: Praeger Publishers, 1998). See also OECD, *Infrastructure Policies for the 1990s* (Paris: OECD, 1993); and Roger W. Vickerman, 'Transport Infrastructure and Region Building in the European Community', *Journal of Common Market Studies*, vol. 32, no. 1, March 1994, pp. 1–24.
2. *The Economist*, 19 August 1995, p. 84.
3. Zygmunt Bauman, *Globalization: The Human Consequences* (Cambridge: Polity Press, 1998); here quoted from Bauman, 'Time and Class: New Dimensions of Stratification', *Sociologisk Rapportserie*, no. 7, Department of Sociology, University of Copenhagen, 1998, pp. 2–3.
4. Paul Virilio, 'Un monde surexposé: fin de l'histoire, ou fin de la géographie?', in *Le Monde Diplomatique*, vol. 44, no. 521, August 1997, p. 17, here quoted from Bauman 'Time and Class'. According to Bauman, the idea of the 'end of geography' was first advanced by Richard O'Brien, in *Global Financial Integration: The End of Geography* (London: Chatham House/Pinter, 1992). See Frances Cairncross, *The Death of Distance: How the Communications Revolution Will Change Our Lives* (Boston, MA: Harvard Business School Press, 1997). See also Linda McDowell, ed., *Undoing Place? A Geographical Reader* (London: Arnold, 1997).
5. *Time*, 3 August 1998.
6. Although dams are not part of transport and communication infrastructure as such, we consider the building of dams to be part of the war of independence from space. Dams typically involve the production of electricity and electricity is one of the most effective ways of freeing industry from localised sources of energy and thus for making industry 'footloose', i.e. independent from space.
7. Peter W. G. Morris and George H. Hough, *The Anatomy of Major Projects: A Study of the Reality of Project Management* (New York: John Wiley & Sons, 1987); Mads Christoffersen, Bent Flyvbjerg and Jørgen Lindgaard Pedersen, 'The Lack of Technology Assessment in Relation to Big Infrastructural Decisions', in *Technology and Democracy: The Use and Impact of Technology Assessment in Europe. Proceedings from the 3rd European Congress on Technology Assessment*, vol. I, Copenhagen: n. p., 4–7 November 1992, pp. 54–75; David Collingridge, *The Management of Scale: Big Organizations, Big Decisions, Big*

Mistakes (London: Routledge, 1992); Joseph S. Szyliowicz and Andrew R. Goetz, 'Getting Realistic About Megaproject Planning: The Case of the New Denver International Airport', *Policy Sciences*, vol. 28, no. 4, 1995, pp. 347–67; Mark Bovens and Paul 't Hart, *Understanding Policy Fiascoes* (New Brunswick, NJ: Transaction Publishers, 1996); Peter Hall, 'Great Planning Disasters Revisited', paper, Bartlett School, London, undated.
8. CNN, *Financial News*, 16 July 1998. For more examples, see Chapters 2–4. See also Elinor Ostrom, Larry Schroeder and Susan Wynne, *Institutional Incentives and Sustainable Development: Infrastructure Policies in Perspective* (Boulder, CO: Westview Press, 1993).
9. *The Economist*, 28 August 1999, p. 47.
10. Edward W. Merrow, *Understanding the Outcomes of Megaprojects: A Quantitative Analysis of Very Large Civilian Projects* (Santa Monica, CA: RAND Corporation, 1988), pp. 2–3.
11. Joanna Gail Salazar, 'Damming the Child of the Ocean: The Three Gorges Project', *Journal of Environment and Development*, vol. 9, no. 2, June 2000, p. 173.
12. Major Projects Association, *Beyond 2000: A Source Book for Major Projects* (Oxford: Major Projects Association, 1994), p. 172; Morris and Hough, *The Anatomy of Major Projects*, p. 214.
13. Ralf C. Buckley, 'How Accurate Are Environmental Impact Predictions?' *Ambio*, vol. 20, nos. 3–4, 1993.
14. Walter Williams, *Honest Numbers and Democracy* (Washington, DC: Georgetown University Press, 1998).
15. Paul C. Huszar, 'Overestimated Benefits and Underestimated Costs: The Case of the Paraguay-Paraná Navigation Study', *Impact Assessment and Project Appraisal*, vol. 16, no. 4, December 1998, p. 303; Philip M. Fearnside, 'The Canadian Feasibility Study of the Three Gorges Dam Proposed for China's Yangtze River: A Grave Embarrassment to the Impact Assessment Profession', *Impact Assessment*, vol. 12, no. 1, spring 1994, pp. 21–57; C. Alvares and R. Billorey, *Damning the Narmada: India's Greatest Planned Environmental Disaster* (Penang, Malaysia: Third World Network and Asia-Pacific People's Environment Network, APPEN, 1988).
16. John F. Kain, 'Deception in Dallas: Strategic Misrepresentation in Rail Transit Promotion and Evaluation', *Journal of the American Planning Association*, vol. 56, no. 2, spring 1990, pp. 184–96; Alan Whitworth and Christopher Cheatham, 'Appraisal Manipulation: Appraisal of the Yonki Dam Hydroelectric Project', *Project Appraisal*, vol. 3, no. 1, March 1988, pp. 13–20; Martin Wachs, 'When Planners Lie with Numbers', *Journal of the American Planning Association*, vol. 55, no. 4, autumn 1989, pp. 476–9; R. Teichroeb, 'Canadian Blessing for Chinese Dam Called "Prostitution"', *Winnipeg Free Press*, 20 September 1990, p. 9.
17. For an empirical case, see Åsa Boholm and Ragnar Löfstedt, 'Issues of Risk, Trust and Knowledge: The Hallandsås Tunnel Case', *Ambio*, vol. 28, no. 6, September 1999, pp. 556–61. For the theoretical argument, see James Bohman, *Public Deliberation: Pluralism, Complexity, and Democracy* (Cambridge, MA: MIT Press, 1996), chap. 3.

18. Brian Doherty, 'Paving the Way: The Rise of Direct Action Against Road-Building and the Changing Character of British Environmentalism', *Political Studies*, vol. 47, no. 2, June 1999, pp. 275–91; Andrea D. Luery, Luis Vega and Jorge Gastelumendi de Rossi, *Sabotage in Santa Valley: The Environmental Implications of Water Mismanagement in a Large-Scale Irrigation Project in Peru* (Norwalk, CT: Technoserve, 1991); Jon Teigland, 'Predictions and Realities: Impacts on Tourism and Recreation from Hydropower and Major Road Developments', *Impact Assessment and Project Appraisal*, vol. 17, no. 1, March 1999, p. 67; 'Svensk webbsida uppmanar till sabotage' (Swedish website is encouraging sabotage) and 'Sabotage för miljoner' (sabotage for millions), *Svensk Vägtidning*, vol. 84, no. 2, 1997, p. 3 and vol. 85, no. 1, 1998, p. 7. One of the authors of the present book has similarly come across sabotage of a large-scale irrigation project in the Kilimanjaro region in Tanzania: see Bent Flyvbjerg, *Making Social Science Matter: Why Social Inquiry Fails and How It Can Succeed Again* (Cambridge: Cambridge University Press, 2001), chap. 10.
19. Ulrich Beck, *Risk Society: Towards a New Modernity* (Thousand Oaks, CA: Sage, 1992); Anthony Giddens, *The Consequences of Modernity* (Stanford, CA: Stanford University Press, 1990); Jane Franklin, ed., *The Politics of Risk Society* (Cambridge: Polity Press, 1998).
20. For an introduction to the literature on risk assessment and management, see Sheldon Krimsky and Dominic Golding, eds., *Social Theories of Risk* (Westport, CT: Praeger, 1992); Ortwin Renn, 'Three Decades of Risk Research: Accomplishments and New Challenges', *Journal of Risk Research*, vol. 1, no. 1, 1998, pp. 49–71. See also Chapter 7 and the three key journals in the field, *Journal of Risk Research*, *Risk Analysis* and *Journal of Risk and Uncertainty*.
21. Silvio O. Funtowicz and Jerome R. Ravetz, 'Three Types of Risk Assessment and the Emergence of Post-normal Science', in Krimsky and Golding, eds., *Social Theories of Risk*, pp. 251–73. See also Carlo Jaeger, Ortwin Renn, Eugene A. Rosa and Thomas Webler, *Risk, Uncertainty and Rational Action* (London: Earthscan, 2001).
22. Ortwin Renn, Thomas Webler and Peter Wiedemann, eds., *Fairness and Competence in Citizen Participation: Evaluating Models for Environmental Discourse* (Dordrecht: Kluwer, 1995); Ortwin Renn, 'A Model for an Analytic-Deliberative Process in Risk Management', *Environmental Science and Technology*, vol. 33, no. 18, September 1999, pp. 3049–55; Thomas Webler and Seth Tuler, 'Fairness and Competence in Citizen Participation: Theoretical Reflections From a Case Study', *Administration and Society*, vol. 32, no. 5, November 2000, pp. 566–95.
23. Adolf G. Gundersen, *The Environmental Promise of Democratic Deliberation* (Madison, WI: University of Wisconsin Press, 1995); Katherine E. Ryan and Lizanne Destefano, eds., *Evaluation As a Democratic Process: Promoting Inclusion, Dialogue, and Deliberation* (San Francisco: Jossey-Bass, 2000); Edward C. Weeks, 'The Practice of Deliberative Democracy: Results From Four Large-Scale Trials', *Public Administration Review*, vol. 60, no. 4, July–August

2000, pp. 360–72. For theories of deliberative democracy, see James Bohman, *Public Deliberation: Pluralism, Complexity, and Democracy* (Cambridge, MA: MIT Press, 1996); Jon Elster, ed., *Deliberative Democracy* (Cambridge: Cambridge University Press, 1998); Stephen MacEdo, ed., *Deliberative Politics: Essays on Democracy and Disagreement* (Oxford: Oxford University Press, 1999); John S. Dryzek, *Deliberative Democracy and Beyond: Liberals, Critics, Contestations* (Oxford: Oxford University Press, 2000).

24. For deliberative approaches to participation based on communicative rationality, see Thomas Webler, ' "Right" Discourse in Citizen Participation: An Evaluative Yardstick' and Frances M. Lynn and Jack D. Kartez, 'The Redemption of Citizen Advisory Committees: A Perspective from Critical Theory', both articles in Renn, Webler and Wiedemann, eds., *Fairness and Competence in Citizen Participation*. For a critique of deliberative approaches and communicative rationality, see Reiner Keller and Angelika Poferl, 'Habermas Fightin' Waste: Problems of Alternative Dispute Resolution in the Risk Society', *Journal of Environmental Policy and Planning*, vol. 2, no. 1, 2000.

25. For a fuller development of this argument, see Bent Flyvbjerg, 'Habermas and Foucault: Thinkers for Civil Society?', *British Journal of Sociology*, vol. 49, no. 2, June 1998, pp. 208–33.

26. For an analysis of approaches based on communicative rationality versus approaches based on power, see Flyvbjerg, *Making Social Science Matter*.

27. Regarding accountability, see P. Day and R. Klein, *Accountabilities* (London: Tavistock, 1987); James G. March and Johan P. Olsen, *Democratic Governance* (New York: Free Press, 1995); Mark Bovens, *The Quest for Responsibility: Accountability and Citizenship in Complex Organisations* (Cambridge University Press, 1998); Fidelma White and Kathryn Hollingsworth, *Audit, Accountability and Government* (Oxford: Clarendon Press, 1999). For further references, see the following chapters.

28. Hall and Taylor distinguish between three different analytical approaches to what is called the 'New Institutionalism': Historical Institutionalism, Rational Choice Institutionalism and Sociological Institutionalism (Peter A. Hall and Rosemary C. R. Taylor, 'Political Science and the Three New Institutionalisms', *Political Studies*, vol. 44, no. 5, 1996, pp. 936–57). In what follows, we draw on the former and latter of these three schools, and especially as they pertain to understanding and transforming the practices and rules of institutional accountability. For more on this topic, see James G. March and Johan P. Olsen, *Rediscovering Institutions: The Organizational Basis of Politics* (New York: Free Press, 1989); Walter W. Powell and Paul J. DiMaggio, eds., *The New Institutionalism in Organizational Analysis* (University of Chicago Press, 1991); W. Richard Scott, *Institutions and Organizations: Theory and Research* (Thousand Oaks, CA: Sage, 1995); and Vivien Lowndes, 'Varieties of New Institutionalism: A Critical Appraisal', *Public Administration*, vol. 74, summer 1996, pp. 181–97.

29. Joe Cummings, *Thailand* (Melbourne: Lonely Planet Publications, 1999), p. 32.

30. Anna Coote, 'Risk and Public Policy: Towards a High-Trust Democracy', in Jane Franklin, ed., *The Politics of Risk Society* (Cambridge: Polity Press, 1998), p. 131.

2 A CALAMITOUS HISTORY OF COST OVERRUN

1. Information from The Channel Tunnel Group, July 1998, kindly made available by Mette K. Skamris, Department of Development and Planning, Aalborg University. According to Eurotunnel, the 1986 Prospectus made provision for a standby loan facility of one billion pounds to provide for such contingencies as delays, additional capital expenditures, etc. The above cost figure does not include this contingency, according to Eurotunnel (correspondence with Eurotunnel, December 1999, authors' archives).
2. Sund & Bælt, *Årsberetning 1999* (Copenhagen: Sund & Bælt Holding, 2000), p. 19.
3. Danish Parliament, 'Bemærkninger til Forslag til Lov om anlæg af fast forbindelse over Øresund', Lovforslag nr. L 178 (Folketinget [Danish Parliament] 1990–91, 2. samling, proposed 2 May 1991), p. 10.
4. Sund & Bælt, *Årsberetning 1999*, p. 20.
5. Sund & Bælt, *Årsberetning 1999*, p. 20; Danish Parliament, 'Redegørelse af 6/12 93 om anlæg af fast forbindelse over Øresund', Fortryk af Folketingets forhandlinger, Folketinget, 7 December 1993 (Copenhagen: sp. 3212–3213); Danish Auditor-General, *Beretning til statsrevisorerne om udviklingen i de økonomiske overslag vedrørende Øresundsforbindelsen* (Copenhagen: Rigsrevisionen (Danish Auditor-General), November 1994), pp. 43–44. Øresundkonsortiet, *Den faste forbindelse over Øresund* (Copenhagen: Øresundkonsortiet, 1994), p. 4.
6. In a response to our description of cost development in the Great Belt and Øresund links, the management of the two links objects to our figures for cost overrun including cost increases that were caused by politically approved changes to project designs, with a view, for instance, to environmental protection. According to the management, such cost increases should not be included in the figures for cost overrun. Also, the management comments that the baseline budget used for calculating cost overrun should not be the budget at the time of decision to build but a later budget estimated after the politically approved design changes and after the project companies had been established and had taken over responsibility for the projects (correspondence with Sund & Bælt Holding, 20 December 1999, authors' archives). If we followed the recommendations of the management, the result would be cost overruns substantially lower than those mentioned in the main text. While we understand why the project management, from their point of view, would prefer not to include in the calculation of cost overrun items and time periods for which they were not responsible, we maintain that the internationally accepted standard for calculating cost overrun is to compare actual costs with costs estimated at the time of decision to build. This is the standard for good reason. First, the information available to those making the decision to build, at the time they make it, is what is relevant when we want to evaluate whether the decision was an informed one or not. Second, this standard for calculating cost overrun

makes it possible to compare the performance of different projects; this would be impossible if we followed the recommendations of the management of the Great Belt and Øresund projects.
7. Swedish Auditor-General, *Infrastrukturinvesteringar: En kostnadsjämförelse mellan plan och utfall i 15 större projekt inom Vägverket och Banverket*, RRV 1994:23 (Stockholm: Avdelningen för Effektivitetsrevision, 1994).
8. Don Pickrell, *Urban Rail Transit Projects: Forecast Versus Actual Ridership and Cost* (Washington, DC: US Department of Transportation, 1990).
9. P. R. Fouracre, R. J. Allport and J. M. Thomson, *The Performance and Impact of Rail Mass Transit in Developing Countries*, TRRL Research Report 278 (Crowthorne: Transport and Road Research Laboratory, 1990).
10. Bent Flyvbjerg, Mette K. Holm and Søren L. Buhl, 'Underestimating Costs in Public Works Projects: Error or Lie?', in *Journal of the American Planning Association*, vol. 68, no. 3, Summer 2002, pp. 279–95. Leonard Merewitz earlier carried out a study aimed at comparing cost overrun in urban rapid transit projects, and especially overrun in the San Francisco Bay Area Rapid Transit (BART) system, with overrun in other types of public works project. This is, to our knowledge, the only other study of cost overrun in transport infrastructure projects with an attempt at establishing statistical significance for its conclusions (Leonard Merewitz, *How Do Urban Rapid Transit Projects Compare in Cost Estimate Experience?* Reprint no. 104, Berkeley: University of California Berkeley, Institute of Urban and Regional Development, 1973; Merewitz, 'Cost Overruns in Public Works', in William A. Niskanen, Arnold C. Harberger, Robert H. Haveman, Ralph Turvey and Richard Zeckhauser, eds., *Benefit–Cost and Policy Analysis*, Chicago: Aldine Publishers, 1973, reprint no. 114, Berkeley: Institute of Urban and Regional Development, University of California, 1973). However, for reasons given elsewhere, Merewitz's study cannot be said to be a true large-N study of transportation infrastructure and its statistical significance is unclear (Flyvbjerg, Skamris Holm and Buhl, 'Underestimating Costs in Public Works Projects: Error or Lie?'). Despite its shortcomings, the approach taken by Merewitz was innovative for its time and in principle pointed in the right direction regarding how to analyse cost overrun in public works projects.
11. The main sources are: Merewitz, *Urban Rapid Transit Projects*; Hall, 'Great Planning Disasters Revisited'; World Bank, *Economic Analysis of Projects: Towards a Results-Oriented Approach to Evaluation*, ECON Report (Washington, DC: forthcoming); R. M. Fraser, 'Compensation for Extra Preliminary and General (P & G) Costs Arising from Delays, Variations and Disruptions: The Palmiet Pumped Storage Scheme', *Tunnelling and Underground Space Technology*, vol. 5, no. 3, 1990; M. M. Dlakwa and M. F. Culpin, 'Reasons for Overrun in Public Sector Construction Projects in Nigeria', *International Journal of Project Management*, vol. 8, no. 4, 1990; Morris and Hough, *The Anatomy of Major Projects: A Study of the Reality of Project Management*; David Arditi, Guzin Tarim Akan and San Gurdamar, 'Cost Overruns in Public Projects', *International Journal of Project Management*, vol. 3, no. 4, 1985; Henry T. Canaday, *Construction Cost Overruns in Electric Utilities: Some Trends and Implications*, Occasional Paper no. 3 (Columbus: National

Regulatory Research Institute, Ohio State University, November 1980); P. D. Henderson, 'Two British Errors: Their Probable Size and Some Possible Lessons', *Oxford Economic Papers*, vol. 29, no. 2, July 1977; Coleman Blake, David Cox and Willard Fraize, *Analysis of Projected Vs. Actual Costs for Nuclear and Coal-Fired Power Plants*, Prepared for the United States Energy Research and Development Administration (McLean, VI: Mitre Corporation, 1976); Department of Energy Study Group (DOE), *North Sea Costs Escalation Study*, Energy Paper no. 8 (London: DOE, 31 December 1975); Maynard M. Hufschmidt and Jacques Gerin, 'Systematic Errors in Cost Estimates for Public Investment Projects', in Julius Margolis, ed., *The Analysis of Public Output* (New York: Columbia University Press, 1970); J. M. Healey, 'Errors in Project Cost Estimates', *Indian Economic Journal*, vol. 12, no. 1, July–September 1964.
12. Hall, 'Great Planning Disasters Revisited', p. 3.
13. Robert Summers, 'Cost Estimates as Predictors of Actual Costs: A Statistical Study of Military Developments', in Thomas Marschak, Thomas K. Glennan and Robert Summers, eds., *Strategy for R&D: Studies in the Microeconomics of Development* (Berlin: Springer-Verlag, 1967), p. 148.
14. Major Projects Association, *Beyond 2000: A Source Book for Major Projects*, p. 165.
15. For more examples of good practice regarding cost estimation and management, see World Bank, *World Development Report 1994: Infrastructure for Development* (Oxford: Oxford University Press, 1994). See also André Blanc, Christian Brossier, Christian Bernardini and Michel Gerard: *Rapport de la Mission sur la Projet de TGV-Est Européen*, Inspection Général des Finances et Conseil Générale des Pont et Chaussées, Paris, July 1996.

3 THE DEMAND FOR MEGAPROJECTS

1. B. Bradshaw and R. Vickerman, eds., *The Channel Tunnel: Transport Studies*, In Focus (Folkestone: The Channel Tunnel Group Ltd, 1993), p. 17.
2. Christoffersen, Flyvbjerg and Pedersen, 'The Lack of Technology Assessment in Relation to Big Infrastructural Decisions', in *Technology and Democracy: The Use and Impact of Technology Assessment in Europe. Proceedings from the 3rd European Congress on Technology Assessment*, vol. I, pp. 54–75.
3. Actual rail passenger traffic was on target when compared to the 1987 forecast, i.e. the forecast at the time of decision to build the Great Belt link. Rail forecasts made after this date turned out to be optimistic: the 1989 forecast of daily rail passenger traffic was 18 per cent higher than actual traffic; the 1991 forecast 65 per cent higher; and the 1994 forecast 32 per cent higher than actual traffic. See A/S Storebæltsforbindelsen, *Øst-vest trafikmodellen: Prognoser for trafikken over Storebæltsbroen og de konkurrerende færgeruter* (Copenhagen: Great Belt Ltd, August 1994), p. 4.
4. Based on month-by-month figures for rail passenger traffic across Great Belt 1997–2000 obtained from DSB, the Danish State Railways, January 2001; and Danish State Railways, *Ånsrapport 2001* at www.dsb.dk, September 2002.
5. Sund & Bælt, *Årsberetning 2001*, p. 6.

6. Danish Parliament, 'Bemærkninger til Forslag til Lov om anlæg af fast forbindelse over Øresund', p. 11.
7. Sund & Bælt, Årsberetning 2001, p. 7; DSB, Årsrapport 2001.
8. Traffic data in the table refer to the opening year. It would have been preferable if forecast and actual traffic could be compared for more years than the opening year. Available data do not allow for this, however.
9. German Federal Ministry for Transport, 'Gesamtwirtschaftliche Bewertung von Verkehrsinfrastrukturinvestitionen für den Bundesverkehrswegeplan', *Schriftenreihe des Bundesministeriums für Verkehr*, Heft 72, Bonn: Bundesminister für Verkehr, 1992.
10. German Federal Ministry for Transport, *Bundesverkehrswegeplan 1992*, Bonn: Bundesminister für Verkehr, 1992.
11. National Audit Office, *Department of Transport, Scottish Department and Welsh Office: Road Planning* (London: HMSO, 1988), p. 2.
12. An earlier evaluation of forty-four UK transport studies showed 'an almost universal overestimation of every item'. I. H. Mackinder and S. E. Evans, *The Predictive Accuracy of British Transport Studies in Urban Areas*, Supplementary Report 699 (Crowthorne: Transport and Road Research Laboratory, 1981), p. 25.
13. *Ibid.*, p. 4.
14. Pickrell, *Urban Rail Transit Projects: Forecast Versus Actual Ridership and Cost*; Pickrell, 'A Desire Named Streetcar: Fantasy and Fact in Rail Transit Planning', *Journal of the American Planning Association*, vol. 58, no. 2, 1992, pp. 158–76. See also Jonathan E. D. Richmond, *New Rail Transit Investments: A Review* (Cambridge, MA: Harvard University, John F. Kennedy School of Government, 1998).
15. Fouracre, Allport and Thomson, *The Performance and Impact of Rail Mass Transit in Developing Countries*, TRRL Research Report no. 278, p. 10.
16. Bent Flyvbjerg and Mette K. Skamris Holm, 'How Accurate are Demand Forecasts in Transport Infrastructure Projects?', paper, forthcoming.
17. To compare historic traffic forecasting with historic electricity forecasting, see Arun Sanghvi and Robert Vernstrom, 'Review and Evaluation of Historic Electricity Forecasting Experience (1960–1985)', *Industry and Energy Department Working Paper, Energy Series Paper*, no. 18, Washington, DC: World Bank, 1989.
18. See also H. O. Stekler, 'Are Economic Forecasts Valuable?' *Journal of Forecasting*, vol. 13, no. 6, 1994, pp. 495–505; Gordon Leitch and J. Ernest Tanner, 'Professional Economic Forecasts: Are They Worth Their Costs?' *Journal of Forecasting*, vol. 14, no. 2, 1995, pp. 143–57; Heidi Winklhofer, Adamantios Diamantopoulos and Stephen F. Witt, 'Forecasting Practice: A Review of the Empirical Literature and an Agenda for Future Research', *International Journal of Forecasting*, vol. 12, no. 2, 1996, pp. 193–221; Mark Garett and Martin Wachs, *Transportation Planning on Trial: The Clean Air Act and Travel Forecasting* (Thousand Oaks, CA: Sage, 1996).
19. M. Gaudry, B. Mandel and W. Rothengatter, 'Linear and Nonlinear Logit Models', *Transportation Research*, part B, vol. 28, no. 6, 1994; B. Mandel, *Schnellverkehr und Modal Split* (Baden Baden: Nomos, 1992).

20. S. P. Huntington and J. S. Nye, Jr, *Global Dilemmas* (Cambridge, MA: Harvard, 1985).
21. For more on the bias of project promoters, see Flyvbjerg, Holm and Buhl, 'Underestimating Costs in Public Works Projects: Error or Lie?', *Journal of the American Planning Association*.

4 SUBSTANCE AND SPIN IN MEGAPROJECT ECONOMICS

1. See also Major Projects Association, *Beyond 2000: A Source Book for Major Projects*, p. 155; *The Economist*, 30 April 1994, pp. 13, 73–4.
2. CNN, *Financial News*, 21 October 1994.
3. *The Economist*, 31 January 1998, p. 74; 24 May 1997, pp. 67–8; 30 April 1994, p. 73; 29 April 1989, p. 73.
4. Danish Parliamentary Auditor's Committee, *Beretning om Storebæltsforbindelsens økonomi*, Beretning 4/97 (Copenhagen: Statsrevisoratet (Auditor's office), 1998), pp. 30–36.
5. The Danish Parliament, Law no. 1233, 27 December 1996.
6. Danish Parliamentary Auditor's Committee, *Beretning om Storebæltsforbindelsens økonomi*, pp. 55–7.
7. Sund & Bælt, *Årsberetning 1999* (Copenhagen: Sund & Bælt Holding, 2000), pp. 20–23.
8. Sund & Bælt, *Årsberetning 2001* (Copenhagen: Sund & Bælt Holding, 2002), p. 27.
9. Sund & Bælt, *Årsberetning 1999*, p. 23.
10. *Ibid.*, p. 22; Sund & Bælt, *Årsberetning 2001*, p. 29.
11. *Ibid.*, p. 24.
12. Danish Parliamentary Auditor's Committee, *Beretning om Storebæltsforbindelsens økonomi*, 1998 p. 58.
13. Danish Parliament, 'Aftale mellem Danmarks regering og Sveriges regering om en fast forbindelse over Øresund', signed 23 March 1991, article 21; Annex to Danish Parliament, 'Forslag til Lov om anlæg af fast forbindelse over Øresund', Lovforslag nr. L 178, proposed 2 May 1991.
14. *Ibid.*
15. Danish Auditor-General, *Beretning til statsrevisorerne om udviklingen i de økonomiske overslag vedrørende Øresundsforbindelsen*, pp. 16–28.
16. *Ibid.*, p. 12.
17. Ministry of Transport, Ministry of Finance and Sund & Bælt Holding, Ltd., *Udredning af økonomien i A/S Øresundsforbindelsen (de danske anlæg)* (Copenhagen: Author, 2002).
18. *Ibid.*, p. 33.
19. Bent Flyvbjerg, 'Economic Risk in Public Works Projects: The Case of Urban Rail', paper, forthcoming.
20. For more on the problem of rent seeking, see Mushtaq H. Khan and K. S. Jomo, eds., *Rents, Rent-Seeking and Economic Development: Theory and Evidence in Asia* (Cambridge: Cambridge University Press, 2000).
21. Swedish Auditor-General, *Infrastrukturinvesteringar: En kostnadsjämförelse mellan plan och utfall i 15 större projekt inom Vägverket och Banverket*, RRV 1994:23.

Notes to pages 42–46　　161

22. *Ibid.*, Annex 2:1.
23. Pickrell, *Urban Rail Transit Projects: Forecast Versus Actual Ridership and Cost*; Richmond, *New Rail Transit Investments: A Review*.
24. Pickrell, *Urban Rail Transport Projects*, p. xi.
25. *Ibid.*, p. xvii. See also Kain, 'Choosing the Wrong Technology: Or How to Spend Billions and Reduce Transit Use', pp. 197–213.
26. Fouracre, Allport and Thomson, *The Performance and Impact of Rail Mass Transit in Developing Countries*, TRRL Research Report no. 278, pp. 7–12.
27. *Ibid.*, p. 14.
28. *Ibid.*, p. 10.
29. The main sources are: Merewitz, 'Cost Overruns in Public Works', in Niskanen *et al.*, eds., *Benefit–Cost and Policy Analysis*; *ibid.*, reprint no. 114; Merewitz, *How Do Urban Rapid Transit Projects Compare in Cost Estimate Experience?*, Reprint no. 104; Hall, *Great Planning Disasters*; World Bank, *Economic Analysis of Projects: Towards a Results-Oriented Approach to Evaluation*, ECON Report; Fraser, 'Compensation for Extra Preliminary and General (P & G) Costs Arising from Delays, Variations and Disruptions: The Palmiet Pumped Storage Scheme'; Dlakwa and Culpin, 'Reasons for Overrun in Public Sector Construction Projects in Nigeria'; Morris and Hough, *The Anatomy of Major Projects: A Study of the Reality of Project Management*; Arditi, Akan and Gurdamar, 'Cost Overruns in Public Projects'; Canaday, *Construction Cost Overruns in Electric Utilities: Some Trends and Implications*; Henderson, 'Two British Errors: Their Probable Size and Some Possible Lessons'; Blake, Cox and Fraize, *Analysis of Projected Vs. Actual Costs for Nuclear and Coal-Fired Power Plants*; Department of Energy Study Group, *North Sea Costs Escalation Study*; Hufschmidt and Gerin, 'Systematic Errors in Cost Estimates for Public Investment Projects', in Margolis, ed., *The Analysis of Public Output*; Healey, 'Errors in Project Cost Estimates'.
30. World Bank, *World Development Report 1994: Infrastructure for Development*, p. 86.
31. Measured as economic rates of return. World Bank, ECON Report, *Economic Analysis of Projects*, pp. 14, 21.
32. World Bank, *World Development Report 1994*, p. 17.
33. World Bank, ECON Report, *Economic Analysis of Projects*, p. 21.
34. Frank P. Davidson and Jean-Claude Huot, 'Management Trends for Major Projects', *Project Appraisal*, vol. 4, no. 3, September 1989, p. 137.
35. Martin Wachs, 'Ethics and Advocacy in Forecasting for Public Policy', *Business and Professional Ethics Journal*, vol. 9, nos. 1 and 2, 1990.
36. Of a sample of forty projects for which it was possible to establish reliable data on both construction cost and traffic development, we were able to identify only five projects with fairly accurate forecasts, defined as actual development falling within an interval of forecast development ±20 per cent. The remaining thirty-five projects fell outside this interval with only one (1) project with forecast cost overestimated and forecast traffic underestimated by more than 20 per cent.
37. See also World Bank, *An Overview of Monitoring and Evaluation in the World Bank*, Report no. 13247, Operations Evaluation Department (Washington, DC: World Bank, 1994); and World Bank, *Evaluation Results*

1992, Operations Evaluation Department (Washington, DC: World Bank, 1994).
38. Pickrell, *Urban Rail Transit Projects: Forecast Versus Actual Ridership and Cost*; Pickrell, 'A Desire Named Streetcar: Fantasy and Fact in Rail Transit Planning', pp. 158–76.
39. Merewitz, 'Cost Overruns in Public Works', in Niskanen *et al.*, eds., *Benefit–Cost and Policy Analysis*, p. 280.
40. Bent Flyvbjerg, 'The Dark Side of Planning: Rationality and *Realrationalität*', in Seymour Mandelbaum, Luigi Mazza and Robert Burchell, eds., *Explorations in Planning Theory* (New Brunswick, NJ: Center for Urban Policy Research Press, 1996); Wachs, 'When Planners Lie with Numbers', pp. 476–479.
41. Wachs, 'Technique Vs. Advocacy in Forecasting: A Study of Rail Rapid Transit', pp. 23–30; Wachs, 'When Planners Lie with Numbers'; Wachs, 'Ethics and Advocacy in Forecasting for Public Policy', pp. 141–57.
42. Wachs, 'Ethics and Advocacy in Forecasting for Public Policy', p. 144.
43. Wachs, 'Technique Vs. Advocacy in Forecasting', p. 28.
44. Wachs, 'Ethics and Advocacy in Forecasting for Public Policy', p. 146; Wachs, 'Technique Vs. Advocacy in Forecasting', p. 28.
45. Flyvbjerg, Holm and Buhl, 'Underestimating Costs in Public Works Projects: Error or Lie?'

5 ENVIRONMENTAL IMPACTS AND RISKS

1. Ben Dipper, Carys Jones and Christopher Wood, 'Monitoring and Post-auditing in Environmental Impact Assessment: A Review', *Journal of Environmental Planning and Management*, vol. 41, no. 6, November 1998, pp. 731, 744.
2. Brundtland Commission, *Our Common Future* (Oxford University Press, 1987); Barry Sadler, *International Study of the Effectiveness of Environmental Assessment*, Interim Report and Discussion Paper (Canadian Environmental Assessment Agency, 1995).
3. Prime Minister's Office, Circular no. 31, Copenhagen: Prime Minister's Office, February 1993 and February 1994.
4. R. Gregory, R. Keeney and D. V. Winterfeldt, 'Adapting the Environmental Impact Statement Process to Inform Decisionmakers', *Journal of Policy Analysis and Management*, vol. 11, no. 1, 1992, p. 61.
5. German Federal Ministry for Transport, *Bundesverkehrswegeplan*.
6. Morten Andersson *et al.*, *Miljøvurdering af transportprojekter: et case studie* (Research Center for Environmental Assessment, Roskilde University, 1994); Andersson and Bo Elling, *Environmental Impact Assessment on Transport Projects in Denmark* (Research Centre for Environmental Assessment, Roskilde University, 1992).
7. T. O'Riordan, 'On Greening of Major Projects', in Major Projects Association, *Major Projects and the Environment*, Proceedings of a Royal Geographical Society/Major Projects Association Conference, Technical Paper, no. 8, Oxford: Major Projects Association, June 1989; Alan Gilpin, *Environmental*

Impact Assessment (EIA): Cutting Edge for the Twenty-First Century (Cambridge: Cambridge University Press, 1995).
8. Christopher Wood, 'Lessons from Comparative Practice', *Built Environment*, vol. 20, no. 4, 1994, pp. 332–344.
9. The Channel Tunnel Group, *The Channel Tunnel Project: Environmental Effects in the UK* (Folkestone: The Channel Tunnel Group Ltd, 1985).
10. Steer Davies and Gleave Ltd, *Turning Trucks into Trains: The Environmental Benefits of the Channel Tunnel* (London: Transport 2000 Ltd., 1987).
11. Bradshaw and Vickerman, eds., *The Channel Tunnel: Transport Studies*, pp. 23–4.
12. Andersson et al., *Miljøvurdering af transportprojekter*.
13. *Ibid*.
14. Joanna R. Treweek, Stewart Thompson, N. Veitch and C. Japp, 'Ecological Assessment of Proposed Road Developments: A Review of Environmental Statements', *Journal of Environmental Planning and Management*, vol. 36, 1993, pp. 295–307.
15. Stewart Thompson, Joanna R. Treweek and D. J. Thurling, 'The Ecological Component of Environmental Impact Assessment: A Critical Review of British Environmental Statements', *Journal of Environmental Planning and Management*, vol. 40, no. 2, 1997, pp. 157–71.
16. Helen J. Byron, Joanna R. Treweek, William R. Sheate and Stewart Thompson, 'Road Developments in the UK: An Analysis of Ecological Assessment in Environmental Impact Statements Produced between 1993 and 1997', *Journal of Environmental Planning and Management*, vol. 43, no. 1, January 2000, pp. 71–97.
17. *Ibid*.
18. BVU, Hensch and Boesefeldt, PLANCO, *Analyse des BVWP-Verfahrens in Methodik und Ablauf* (Bonn: Bundesminister für Verkehr [German Federal Ministry for Transport], 1997).
19. EU Commission, *Guidelines for the Construction of a Transeuropean Transport Network*, EU Bulletin L228 (Brussels: EU Commission, 1996).
20. Graham Wood, 'Post-development Auditing of EIA Predictive Techniques: A Spatial Analytical Approach', *Journal of Environmental Planning and Management*, vol. 42, no. 5, September 1999, pp. 684, 687; Dipper, Jones and Wood, 'Monitoring and Post-auditing in Environmental Impact Assessment: A Review', p. 744.
21. *Ibid*.
22. Ralf C. Buckley, 'Auditing the Precision and Accuracy of Environmental Impact Predictions in Australia', *Environmental Monitoring and Assessment*, vol. 18, 1991; Buckley, 'How Accurate are Environmental Impact Predictions?', *Ambio*, vol. 20, nos. 3–4, 1991.
23. Paul J. Culhane, 'The Precision and Accuracy of U.S. Environmental Impact Statements', *Environmental Monitoring and Assessment*, vol. 8, no. 3, 1987.
24. See also Ronald Bisset, 'Post Development Audits to Investigate the Accuracy of Environmental Impact Predictions', *Zeitschrift für Umweltpolitik*, vol. 7, no. 4, 1984, pp. 463–84; Larry W. Canter, 'Impact Prediction Auditing', *The Environmental Professional*, vol. 7, no. 3, 1985, pp. 255–64;

United Nations Economic Commission for Europe, *Post-project Analysis in Environmental Impact Assessment* (New York: United Nations Publications, 1990); John Glasson, 'Life After The Decision: The Importance of Monitoring in EIA', *Built Environment*, vol. 20, no. 4, 1994, pp. 309–20; Ashley Bird and Riki Therivel, 'Post-auditing of Environmental Impact Statements Using Data Held in Public Registers of Environmental Information', *Project Appraisal*, vol. 11, no. 2, June 1996, pp. 105–16.
25. Buckley, 'Auditing the Precision and Accuracy of Environmental Impact Predictions in Australia', p. 20. See also the critical remarks of Alho, who argues that the source of Buckley's pessimistic view on the accuracy of prediction data is mainly his statistical approach while the underlying data allow for a more positive judgement; Juha M. Alho, 'The Accuracy of Environmental Impact Assessments: Skew Prediction Errors', *Ambio*, vol. 21, no. 4, 1992, pp. 322–3. See also Buckley's refutation of Alho's argument: Buckley, 'Response to Comment by J. M. Alho', *Ambio*, vol. 21, no. 4, 1992, pp. 323–4.
26. Christopher Wood, Ben Dipper and Carys Jones, 'Auditing the Assessment of the Environmental Impacts of Planning Projects', *Journal of Environmental Planning and Management*, vol. 43, no. 1, January 2000.
27. Kim Lynge Nielsen, 'Environmental Appraisal of Large Scale Transport Infrastructure Projects', unpublished Ph.D dissertation (Aalborg: Aalborg University, Department of Development and Planning, 2000).
28. Sund & Bælt, *Storebælt og miljøet* (Copenhagen: Sund & Bælt Holding, 1999), pp. 14–16; Sund & Bælt, 'Omkostninger og fordele ved miljøprioritering', unpublished paper (Copenhagen: Sund & Bælt, 14 April 2000).
29. Sund & Bælt, *Storebælt og miljøet*, p. 7.
30. Sund & Bælt, 'Omkostninger og fordele ved miljøprioritering', p. 2.
31. Sund & Bælt, *Storebælt og miljøet*, pp. 4, 31–6. Ministry of Environment and Energy, Ministry of Transport and the Control and Steering Group for the Øresund Link, *Afslutningsrapport om miljøet og Øresundsforbindelsens kyst-til-kyst anlæg, 11, halvårsrapport* (Copenhagen: Authors, 2001).
32. *Ibid.*, p. 4. The final documentation of impacts on marine flora and fauna was not available at the time of writing this book.
33. *Ibid.*, p. 64.
34. *Ibid.*
35. Henrik Duer, 'Øst–Vest trafikkens energiforbrug næsten uændret', *Transportrådets Nyhedsbrev*, no. 3, September 1999, p. 10.
36. *Ibid.*, p. 9.
37. Henrik Duer, 'Energiforbrug i Øst–Vest trafikken', unpublished paper (Lyngby: Cowi, 14 September 1999).
38. See also David R. McCallum, 'Follow-Up to Environmental Impact Assessment: Learning from the Canadian Government Experience', *Environmental Monitoring and Assessment*, vol. 8, no. 3, 1987, pp. 199–215; Edwin D. Pentecost, 'Managing Large-Scale Environmental Impact Assessments', *Impact Assessment Bulletin*, vol. 9, no. 4, 1991, pp. 35–40; Robert B. Beattie, 'Everything You Already Know About EIA (But Don't Often Admit)', *Environmental Impact Assessment Review*, vol. 15, no. 2, March 1995, pp. 109–14;

David P. Lawrence, 'Quality and Effectiveness of Environmental Impact Assessments: Lessons and Insights from Ten Assessments in Canada', *Project Appraisal*, vol. 12, no. 4, December 1997, pp. 219–32; Joseph S. Szyliowicz, 'Decision Making for Sustainable Development: Towards a New Paradigm', paper prepared for IPSA World Congress, Seoul, Korea, August 1997; Wen-Shyan Leu, W. P. Williams and A. W. Bark, 'Evaluation of Environmental Impact Assessment in Three Southeast Asian Nations', *Project Appraisal*, vol. 12, no. 2, June 1997, pp. 89–100; John Glasson et al., 'EIA – Learning from Experience: Changes in the Quality of Environmental Impact Statements for UK Planning Projects', *Journal of Environmental Planning and Management*, vol. 40, no. 4, 1997, pp. 451–64; Wood, Dipper and Jones, 'Auditing the Assessment of the Environmental Impacts of Planning Projects', pp. 23–47.

39. See also Timothy O'Riordan, *Major Projects and the Environmental Movement*, Major Projects Association Technical Paper, no. 5 (Oxford: Major Projects Association, April 1988); Paul Charest, 'Aboriginal Alternatives to Megaprojects and Their Environmental and Social Impacts', *Impact Assessment*, vol. 13, no. 4, 1995, pp. 371–86; Yuen Ching Ng and W. R. Sheate, 'Environmental Impact Assessment of Airport Development Proposals in the United Kingdom and Hong Kong: Who Should Participate?' *Project Appraisal*, vol. 12, no. 1, March 1997, pp. 11–24; Anne Shepherd and Christi Bowler, 'Beyond the Requirements: Improving Public Participation in EIA', *Journal of Environmental Planning and Management*, vol. 40, no. 6, 1997, pp. 725–38; Hugh Ward, 'Citizens' Juries and Valuing the Environment: A Proposal', *Environmental Politics*, vol. 8, no. 2, summer 1999, pp. 75–96; John Randolph and Michael Bauer, 'Improving Environmental Decision-Making Through Collaborative Methods', *Policy Studies Review*, vol. 16, no. 3–4, fall–winter 1999, pp. 168–91; Juan R. Palerm, 'An Empirical-Theoretical Analysis Framework for Public Participation in Environmental Impact Assessment', *Journal of Environmental Planning and Management*, vol. 43, no. 5, September 2000, pp. 581–600.

40. On the issue of risk assessment in environmental impact assessment, see M. Carlota Arquiaga, Larry W. Canter and Deborah Imel Nelson, 'Risk Assessment Principles in Environmental Impact Studies', *The Environmental Professional*, vol. 14, no. 3, 1992, pp. 204–19; Audrey M. Armour, 'Risk Assessment in Environmental Policymaking', *Policy Studies Review*, vol. 12, nos. 3–4, autumn–winter 1993, pp. 178–196; Joshua Lipton et al., 'A Paradigm for Ecological Risk Assessment', *Environmental Management*, vol. 17, no. 1, 1993, pp. 1–5; National Research Council, *Issues in Risk Assessment: A Paradigm for Ecological Risk Assessment* (Washington, DC: National Academy Press, 1993); O. A. Sankoh, 'An Evaluation of the Analysis of Ecological Risks Method in Environmental Impact Assessment', *Environmental Impact Assessment Review*, vol. 16, no. 3, May 1996, pp. 183–8.

41. See also Maria Rosário Partidário, 'Strategic Environmental Assessment: Key Issues Emerging from Recent Practice', *Environmental Impact Assessment Review*, vol. 16, no. 1, January 1996, pp. 31–55; Tim Richardson, 'The Trans-European Transport Network: Environmental Policy Integration in the

European Union', *European Urban and Regional Studies*, vol. 4, no. 4, 1997, pp. 333–46.
42. Funtowicz and Ravetz, 'Three Types of Risk Assessment and the Emergence of Post-normal Science', in Krimsky and Golding, eds., *Social Theories of Risk*, pp. 270–71.
43. For more on environmental monitoring and auditing, see Ralf C. Buckley, 'Environmental Audit: Review and Guidelines', *Environment and Planning Law Journal*, vol. 7, no. 2, June 1990, pp. 127–41; Larry W. Canter, 'The Role of Environmental Monitoring in Responsible Project Management', *The Environmental Professional*, vol. 15, no. 1, 1993, pp. 76–87; Neil Gunningham, 'Environmental Auditing: Who Audits the Auditors?', *Environmental and Planning Law Journal*, vol. 10, no. 4, 1993, pp. 229–38; Helmut Karl, 'Better Environmental Future in Europe Through Environmental Auditing?', *Environmental Management*, vol. 18, no. 4, 1994, pp. 617–21; Dixon Thompson and Melvin J. Wilson, 'Environmental Auditing: Theory and Applications', *Environmental Management*, vol. 18, no. 4, 1994, pp. 605–15; Peter Bein and Mike Kawczynski, 'Environmental Accounting Applied to Greater Vancouver Transportation System Planning', paper presented to the 76th Annual Meeting of the Transportation Research Board (Washington, DC: Transportation Research Board, January 1997).
44. Wood, Dipper and Jones, 'Auditing the Assessment of the Environmental Impacts of Planning Projects', p. 46; Dipper, Jones and Wood, 'Monitoring and Post-auditing in Environmental Impact Assessment: A Review', p. 744.

6 REGIONAL AND ECONOMIC GROWTH EFFECTS

1. A number of recent studies in the USA, Germany, Japan, Mexico, Sweden and the UK claim to have found that infrastructure investments significantly contribute to economic growth by reducing production costs. See D. A. Aschauer, 'Is Public Expenditure Productive?', *Journal of Monetary Economics*, vol. 23, 1989, pp. 177–200 and Aschauer, 'Public Infrastructure Investment: A Bridge to Productivity Growth?', *Public Policy Brief* 4 (Annandale-on-Hudson, NY: Bard College, Jerome Levy Economic Institute, 1993). These studies have been subject to criticism as they have not shown whether it is investments in infrastructure that cause economic growth or vice versa, and as other factors that may cause both economic growth and investments in infrastructure have not been taken into account. See World Bank, *World Development Report 1994: Infrastructure for Development*, pp. 14–15. See also Arturo Israel, 'Issues for Infrastructure Management in the 1990s', *World Bank Discussion Papers*, no. 171, Washington, DC: World Bank, 1992. As shown by Winston, the results of the 1989 study by Aschauer are implausible as they suggest a cost–benefit ratio of 10:1 for investments in infrastructure, and such returns are implausible and have never been recorded through *ex post* evaluations; Clifford Winston, 'Efficient Transportation Infrastructure Policy', *Journal of Economic Perspective*, vol. 5, no. 1, winter 1991, pp. 113–27. It has been argued that Aschauer's study simply shows that the time pattern of productivity and public investment growth are similar – both

rising in the 1950s and 1960s, and both falling in the 1970s and 1980s – and that this correlation generates grossly inflated estimates of the return to public infrastructure investment; see Charles L. Schultze, 'The Federal Budget and the Nation's Economic Health', in Henry Aaron, ed., *Setting National Priorities: Policy for the Nineties* (Washington, DC: Brookings Institution, 1990). For a comprehensive review of the debate and critique of the 'macroeconomic approach' introduced by Aschauer, see Edward M. Gramlich, 'Infrastructure Investment: A Review Essay', *Journal of Economic Literature*, vol. 32, September 1994, pp. 1176–96; see also Damian J. Kulash, 'Economic Returns from Transportation Investment', *Transportation Quarterly*, vol. 51, no. 3, summer 1997, pp. 8–19.
2. Khan and Jomo, eds., *Rents, Rent-Seeking and Economic Development: Theory and Evidence in Asia*; Peter Boothroyd *et al.*, 'The Need for Retrospective Impact Assessment: The Megaprojects Example', *Impact Assessment*, vol. 13, no. 3, September 1995, pp. 253–71.
3. See Herbert Mohring, 'Maximising, Measuring and *Not* Double Counting Transportation-Improvement Benefits: A Primer on Closed- and Open-Economy Cost–Benefit Analysis', *Transportation Research*, B, vol. 27B, no. 6, 1993, pp. 413–24 and Curt Carnemark, Jaime Biderman and David Bovet, *The Economic Analysis of Rural Road Projects*, World Bank Staff Working Papers, no. 214 (Washington, DC: World Bank, 1984).
4. A. J. Venables and M. Gasiourek, 'Evaluating Regional Infrastructure: A Computable Approach', paper, London School of Economics, 1996.
5. Roger W. Vickerman, 'The Channel Tunnel and Regional Development: A Critique of an Infrastructure-Led Growth Project', *Project Appraisal*, vol. 2, no. 1, March 1987, p. 38.
6. *Ibid.*, p. 40.
7. *Ibid.*
8. Roger Vickerman, 'Long Term Impacts of the Channel Tunnel: Methodology and Evidence', paper for International Research Seminar on the Regional Development Impacts of the Øresund Bridge, Copenhagen, 28–29 November 1999, pp. 11–12.
9. Danish Transport Council, 'Regionaløkonomiske effekter af investeringer i trafikinfrastruktur: Notat til Trafikministeren', in Danish Transport Council, *Fire baggrundsnotater til trafik 2005*, Notat no. 93:09 (Copenhagen; Transportrådet (Danish Transport Council), December 1993), p. 7. Two other reports issued by the Danish Transport Council contain further Danish and international evidence of relevance to the subject of this chapter: see Anne-Mette Hjalager, '*Transportinfrastruktur og regional udvikling: Danske undersøgelser*', Transportrådet, Notat no. 93:07 (Copenhagen: Transportrådet, December 1993); and Bjarne Madsen, Chris Jensen-Butler and Thomas Bue Bjørner, *Transportinfrastruktur og regional udvikling: Udenlandske undersøgelser*, Transportrådet, Notat no. 93:06 (Copenhagen: December 1993).
10. Christian Wichmann Matthiesen and Åke E. Andersson, *Øresundsregionen: Kreativitet, Integration, Vækst* (Copenhagen: Munksgaard, 1993) and Den svenska Öresundsdelegationens Plan- och Miljögrupp, *Öresundsförbindelser: Planerings- och miljöfrågor*, DsK 1978:6, Stockholm: Den svenska Öresundsdelegation, 1978.

11. Matthiesen and Andersson, *Øresundsregionen*.
12. Henry Tordenström, *Trafik- och näringslivseffekter av broar och andra trafiklänkar: Internationella erfarenheter av genomförda projekt* (Malmö: Stadskontoret, 1991).
13. Koschatzky, 'A River is a River: Cross-Border Networking Between Baden and Alsace', pp. 429–49.
14. Hjalager, *Transportinfrastruktur*, p. 6; David Hurdle, 'Does Transport Investment Stimulate Economic Activity?', *The Planner*, vol. 78, no. 9, May 1992. See also Geir Olav Ryntveit and Jon Inge Lian, *Ringvirkninger av endret veitilgjenglighet*, TØI-rapport 168, Oslo, n.p., 1993; Department of the Environment, Transport and the Regions, *Transport and the Economy* (London: Standing Advisory Committee on Trunk Road Assessment, 1999); Roger Vickerman, 'Economic Impacts of Large Transport Infrastructure Projects', paper for Symposium on Effects of a Maglev-Train Schiphol–Groningen, Groningen, 17 November 2000.
15. See e.g. *ibid.*, and Hurdle, 'Does Transport Investment Stimulate Economic Activity?'
16. J. S. Dodgson, 'Motorway Investment, Industrial Transport Costs, and Sub-regional Growth: A Case Study of the M62', *Regional Studies*, vol. 8, 1973, pp. 75–91. See also Robert Cervero and John Landis, 'Assessing the Impacts of Urban Rail Transit on Local Real Estate Markets Using Quasi-experimental Comparisons', *Transportation Research*, Part A, vol. 27A, no. 1, 1993, pp. 13–22; Sarah Stewart Wells and Bruce G. Hutchinson, 'Impact of Commuter-Rail Services in Toronto Region', *Journal of Transportation Engineering*, vol. 122, no. 4, July–August 1996, pp. 270–5.
17. World Bank, *World Development Report*; Transportrådet, 1993, 'Regionaløkonomiske effekter of investeringer...', and Hjalager, *Transportinfrastruktur*. See also Dieter Biehl, *The Contribution of Infrastructure to Regional Development*, Final Report, Infrastructure Study Group (Luxembourg: Commission of the European Communities, 1986).
18. J. Morlok, 'Baden Airpark and its Importance for the High Technology Region of Karlsruhe', paper presented at Seminar on Soft Factors in Spatial Dynamics, University of Karlsruhe, February 1998.

7 DEALING WITH RISK

1. See also David Lewis, 'The Future of Forecasting: Risk Analysis as a Philosophy of Transportation Planning', *TR News*, no. 177, March 1995, pp. 3–9; for more on the problem of predicting performance; and Steven Schnaars, *Megamistakes: Forecasting and the Myth of Technological Change* (New York: Free Press, 1989).
2. In this chapter the main focus is on economic and financial risks. See Chapter 5 for environmental risks.
3. For an introduction to risk analysis, see David B. Hertz and Howard Thomas, *Risk Analysis and Its Applications* (New York: John Wiley & Sons Ltd, 1984); Dale Cooper and Chris Chapman, *Risk Analysis for Large Projects: Models, Methods, and Cases* (New York: John Wiley & Sons, 1987); M. Granger

Morgan, Max Henrion and Mitchell Small, *Uncertainty: A Guide to Dealing with Uncertainty in Quantitative Risk and Policy Analysis* (Cambridge University Press, 1990); Savvakis Savvides, 'Risk Analysis in Investment Appraisal', *Project Appraisal*, vol. 9, no. 1, March 1994, pp. 3–18; Jaeger, Renn, Rosa and Webler, *Risk, Uncertainty and Rational Action*.
4. For more theoretical approaches to the risk issue, see Kenneth J. Arrow and Robert C. Lind, 'Uncertainty and the Evaluation of Public Investment Decisions', *American Economic Review*, vol. 60, 1970, pp. 364–378 and Krimsky and Golding, eds., *Social Theories of Risk*.
5. Quoted from 'Under Water Over Budget', *The Economist*, 7 October 1989, pp. 37–8. The Maître d'Oeuvre was an organisation established to monitor project planning and implementation for the Channel tunnel. It was established in 1985, and until 1988 it represented the owners. From 1988 it reverted to an impartial position. See Major Projects Association, *Beyond 2000: A Source Book for Major Projects*, pp. 151–3.
6. *The Economist*, p. 37.
7. Danish Parliamentary Committee on Public Works, 'Udvalgets spørgsmål til ministeren for offentlige arbejder og dennes besvarelse heraf', annex to 'Betænkning over Forslag til lov om anlæg af fast forbindelse over Storebælt', report from the Parliamentary Committee on Public Works, 19 May 1987, p. 21.
8. Mette K. Skamris and Bent Flyvbjerg, 'Inaccuracy of Traffic Forecasts and Cost Estimates on Large Transport Projects', *Transport Policy*, vol. 4, no. 3, 1997, pp. 141–6.
9. Operational risk analyses have been carried out, e.g. analyses of collision risks and weather risks. The types of risk analysis addressed in this book are mainly financial, economic and environmental risk analyses.
10. Økonomigruppen (The Economy Group), 'Notat om Øresundsforbindelsens driftsøkonomiske rentabilitet', Copenhagen: Ministry of Transport, Ministry of Finance, Road Directorate, Danish State Railways, 2 October 1990), p. 6.
11. Danish Auditor-General, *Beretning til statsrevisorerne om udviklingen i de økonomiske overslag vedrørende Øresundsforbindelsen*, p. 12.
12. *Ibid.*, pp. 11–12.
13. *Ibid.*, p. 12.
14. *Ibid.* p. 11.
15. *Ibid.*, p. 13.
16. World Bank, *Economic Analysis of Projects: Towards a Results-Oriented Approach to Evaluation*, ECON Report, pp. 29–30. See also David Lewis, 'The Future of Forecasting'.
17. Blanc, Brossier, Bernardini and Gerard, *Rapport de la Mission sur la Projet de TGV-Est Européen*. The construction of high-speed railway lines (TGV) in France has been characterised by only small cost overruns. The cost overruns of TGV Sud-Est and TGV Atlantique were only 0.86 per cent, that of TGV Nord 2.6 per cent while the highest cost overrun registered so far was for TGV Rhône-Alpes.
18. Morris and Hough, *The Anatomy of Major Projects: A Study of the Reality of Project Management* (New York: Wiley & Sons, 1987), p. 13.

19. World Bank, *World Development Report 1994: Infrastructure for Development*; Peter K. Nevitt, *Project Financing* (London: Euromoney Publications, 1983); Carl R. Beidleman, Donna Fletcher and David Veshosky, 'On Allocating Risk: The Essence of Project Finance', *Sloan Management Review*, vol. 31, no. 3, spring 1990.
20. Robert C. Lind, ed., *Discounting for Time and Risk in Energy Policy* (Baltimore: Johns Hopkins University Press, 1982).
21. *Ibid.*, and W. F. Sharp, *Investments* (Englewood Cliffs, NJ: Prentice Hall, 1978).
22. Lind, *Discounting for Time*, and Department of Finance, Australia: 'The Choice of Discount Rate for Evaluating Public Sector Investment Projects: A Discussion Paper' (Canberra: Department of Finance, November 1987).
23. Data supplied by the European Bank for Reconstruction and Development.
24. Personal communication, authors' archives.
25. World Bank, *World Development Report 1994*, pp. 91, 100.
26. Personal communication with Great Belt Ltd.
27. World Bank, *Economic Analysis of Projects*, p. 22.
28. *Ibid.*, pp. ii, 22.
29. For theoretical approaches to risk management, see James G. March and Zur Shapira, 'Managerial Perspectives on Risk and Risk Taking', *Management Science*, vol. 33, no. 11, November 1987, pp. 1404–18; Keith Redhead and Steward Hughes, *Financial Risk Management* (Aldershot: Gower, 1988); Jean Couillard, 'The Role of Project Risk in Determining Project Management Approach', *Project Management Journal*, vol. 26, no. 4, 1995, pp. 3–15; Kenneth J. Arrow, 'The Theory of Risk-Bearing: Small and Great Risks', *Journal of Risk and Uncertainty*, vol. 12, nos. 2–3, 1996, pp. 103–11. See also T. Webler, H. Rakel and R. J. S. Ross, 'A Critical Theoretic Look at Technical Risk Analysis', *Industrial Crisis Quarterly*, vol. 6, 1992, pp. 23–38. For practical approaches to risk management, see Council of Standards Australia, and Council of Standards New Zealand, *Risk Management*, AS/NZS 4360:1995 (Homebush: Standards Australia, and Wellington: Standards New Zealand, 1995); Tim Boyce, *Commercial Risk Management: How to Identify, Mitigate and Avoid the Principal Risks in Any Project* (London: Thorogood, 1995).
30. Arrow and Lind, 'Uncertainty and the Evelution of Public Investment Decisions'.
31. See also J. G. Morone and E. J. Woodhouse, *Averting Catastrophe: Strategies for Regulating Risky Technologies* (Berkeley: University of California Press, 1986), and W. T. Singleton and Jan Hovden, eds., *Risk and Decisions* (New York: John Wiley & Sons, 1987).

8 CONVENTIONAL MEGAPROJECT DEVELOPMENT

1. For a review of the approach to development of the Channel tunnel project, see Chapter 9.
2. For other critiques of the conventional approach to megaproject development and appraisal, see Paul R. Schulman, *Large-Scale Policy Making* (New York: Elsevier, 1980); O. P. Kharbanda and E. A. Stallworthy, *How to Learn from*

Project Disasters: True-Life Stories with a Moral for Management (Aldershot: Gower, 1983); Victor Bignell and Joyce Fortune, *Understanding Systems Failures* (Manchester University Press, 1984); Schnaars, *Megamistakes: Forecasting and the Myth of Technological Change*; David Collingridge, *The Management of Scale: Big Organizations, Big Decisions, Big Mistakes* (London: Routledge, 1992); Szyliowicz and Goetz, 'Getting Realistic About Megaproject Planning: The Case of the New Denver International Airport', pp. 347–67; Ingemar Ahlstrand, *Från särintresse till allmänintresse: Om beslutsunderlagets centrala betydelse med exempel från Öresundsbron och Dennispaketet* (Stockholm: SNS Förlag, 1995); Martin H. Krieger, ed., 'Big Decisions, Big Projects, Big Plans', book manuscript (Los Angeles, CA: University of Southern California, School of Urban Planning and Development, undated).

3. The over-commitment of resources and political prestige at an early stage, together with an early over-focus on technical solutions (as opposed to a focus on policy objectives: see next point in main text), is a classical cause of societies getting locked into under-performing projects. See Elliot J. Feldman, 'Patterns of Failure in Government Megaprojects: Economics, Politics, and Participation in Industrial Democracies', in Samuel P. Huntington and Joseph S. Nye, Jr, eds., *Global Dilemmas* (Cambridge, MA: Harvard University, Center for International Affairs, and Lanham, MD: University Press of America, 1985); Gerald M. Steinberg, 'Large-Scale National Projects as Political Symbols: The Case of Israel', *Comparative Politics*, vol. 19, no. 3, April 1987, pp. 331–46; James Maxwell *et al.*, 'Locked on Course: Hydro-Quebec's Commitment to Mega-projects', *Environmental Impact Assessment Review*, vol. 17, no. 1, 1997, pp. 19–38.

4. David Collingridge, 'Undemocratic Technology: Big Organization, Big Governments, Big Projects and Big Mistakes', paper presented at Technological Strategies for the New Europe, International Seminar, UNESCO, undated.

5. Claus Hedegaard Sørensen, *Slår bro fra kyst til kyst: en analyse af Socialdemokratiet og Øresundsforbindelsen* (Århus: Hovedland, 1993); Gunnar Falkemark, *Öresundsbron: Hur de avgörande besluten togs*, Rapport (Stockholm: Naturskyddsföreningen, 1992); Ahlstrand, *Från särintresse till allmänintresse*, chap. 2.

6. See also John F. L. Ross, 'When Co-operation Divides: Øresund, the Channel Tunnel and the New Politics of European Transport', *Journal of European Public Policy*, vol. 2, no. 1, March 1995, pp. 115–46.

9 LESSONS OF PRIVATISATION

1. See for example Barry Eichengreen, *Financing Infrastructure in Developing Countries: Lessons From the Railway Age*, Policy Research Working Paper no. 1379 (Washington, DC: World Bank, 1994).

2. Vincent Wright, ed., *Privatization in Western Europe: Pressures, Problems and Paradoxes* (London: Pinter Publishers, 1994); Paul Seidenstat, 'Privatization: Trends, Interplay of Forces, and Lessons Learned', *Policy Studies Journal*, vol. 24, no. 3, 1996, pp. 464–77.

3. Elizabeth E. Bailey and Janet Rothenberg Pack, eds., *The Political Economy of Privatization and Deregulation* (Aldershot: Edward Elgar, 1995). On privatisation in the transport sector, see José A. Gómez-Ibáñez and John

R. Meyer, *Going Private: The International Experience with Transport Privatization* (Washington, DC: Brookings Institution, 1993); John Archer and Tracy Dunleavy, 'Rethinking Transportation Financing Policies', *Transportation Quarterly*, vol. 50, no. 3, summer 1996, pp. 131–44; Jonathan E. D. Richmond, *The Private Provision of Public Transport* (Cambridge, MA: Harvard University Taubman Center for State and Local Government, 2001).

4. From the very large literature on these and other issues related to privatisation, see Kate Ascher, *The Politics of Privatization: Contracting Out Public Services* (New York: St Martin's Press, 1987); Steve H. Hanke, ed., *Prospects for Privatization* (New York: Academy of Political Science, 1987); Ronald C. Moe, 'Exploring the Limits of Privatization', *Public Administration Review*, vol. 47, no. 6, 1987, pp. 453–60; Sheila B. Kamerman and Alfred J. Kahn, eds., *Privatization and the Welfare State* (Princeton, NJ: Princeton University Press, 1989); and Bruce A. Wallin, 'The Need for a Privatization Process: Lessons from Development and Implementation', *Public Administration Review*, vol. 51, no. 1, 1997, pp. 11–20; Elliott D. Sclar, *You Don't Always Get What You Pay For: The Economics of Privatization* (Ithaca, NY: Cornell University Press, 2000).

5. Project finance can be defined as finance raised on the merits of the project itself. The project becomes the source of the repayment, i.e. once it is completed, its cash flow is used to repay the loans. The financing of a project is said to be non-recourse when lenders are repaid only from the cash flow generated by the project or, in the event of complete failure, from the value of the project's assets. Lenders may also have limited recourse to the assets of a parent company sponsoring a project.

6. S. C. McCarthy and R. L. K. Tiong, 'Financial and Contractual Aspects of Build-Operate-Transfer Projects', *International Journal of Project Management*, vol. 9, no. 4, November 1991, pp. 222–7; Philip Blackshaw, John Flora and Richard Scurfield, 'Motorways by BOT: Political Dogma or Economic Rationality?', in *Financing Transport Infrastructure* (London: PTRC Publications, 1992), pp. 1–13; Jens Vestgaard, 'International Experience with Contractor Driven BOT Concessions. Case: A Contractor Promotes a BOT Project with MOT in a Developing Country', pp. 419–29, in Harry Lahrmann and Leif Hald Pedersen, eds., *Trafikdage på Aalborg Universitet '96* (Aalborg: Danish Transport Council and Aalborg University Transport Research Group, 1996).

7. IFC, The International Finance Corporation, is a member of the World Bank Group. It invests in commercial ventures, normally without sovereign guarantees. Gary Bond and Laurence Carter, 'Financing Private Infrastructure Projects: Emerging Trends from IFC's Experience', IFC Discussion Paper Number 23, Washington, DC, 1994, p. x; World Bank, *World Development Report, 1994: Infrastructure for Development*, p. 100.

8. Peter Moles and Geoffrey Williams, 'Privately Funded Infrastructure in the UK: Participants' Risk in the Skye Bridge Project', *Transport Policy*, vol. 2, no. 2, 1995, pp. 129–34.

9. Stephen Lockwood, Ravindra Verma and Michael Schneider, 'Public–Private Partnerships in Toll Road Development: An Overview of Global Practices', *Transportation Quarterly*, vol. 54, no. 2, spring 2000, pp. 77–91.

10. World Bank, *World Development Report 1994* and Gordon Mills, 'Commercial Funding of Transport Infrastructure: Lessons from Some Australian Cases', *Journal of Transport Economics and Policy*, vol. 25, no. 3, September 1991, pp. 279–98.
11. Moles and Williams, 'Privately Funded Infrastructure in the UK: Participants' Risk in the Skye Bridge Project'.
12. Major Projects Association, *Beyond 2000: A Source Book for Major Projects*, p. 166.
13. See Tony Ridley, 'The Chunnel', *Land and Infrastructure*, undated, pp. 14–19, and Major Projects Association, *Beyond 2000*, pp. 161–74.
14. Ridley, 'The Chunnel', and Major Projects Association, *Beyond 2000*.
15. See e.g. Asian Development Bank, *Developing Best Practices for Promoting Private Sector Investment in Infrastructure: Roads* (Manila, Philippines, 2000).
16. See National Audit Office, *The Private Finance Initiative: The First Four Design, Build, Finance and Operate Roads Contracts*, Report by the Comptroller and Auditor-General, London, 28 January 1998.
17. See also Mills, 'Commercial Funding'.
18. Lockwood, Verma and Schneider, 'Public–Private Partnerships in Toll Road Development', pp. 77–91. Regarding private financing of railways, we refer the reader to the extensive British experience; see for instance C. A. Nash, 'Developments in Transport Policy: Rail Privatisation in Britain', *Journal of Transport Economics and Policy*, vol. 27, no. 3, 1993, pp. 317–22; and John Preston, 'The Economics of British Rail Privatization: An Assessment', *Transport Reviews*, vol. 16, no. 1, 1996, pp. 1–21. See also Gómez-Ibáñez and Meyer, *Going Private*, and Ravi Ramamurti, 'Testing the Limits of Privatization: Argentine Railroads', *World Development*, vol. 25, no. 12, December 1997, pp. 1973–93.
19. This section on toll roads in France and Spain is based mainly on Gómez-Ibáñez and Meyer, *Going Private*. See also Alain Fayard, 'Overview of the Scope and Limitations of Public–Private Partnerships', Seminar of Public–Private Partnerships (PPPs) in Transport Infrastructure Financing, European Conference of Ministers of Transport, Paris, 1999. Regarding the French experience with private financing, see also Claude Martinand, ed., *Private Financing of Public Infrastructure: The French Experience* (Paris: Ministry of Public Works, Transportation and Tourism, 1994); and G. Charpentier, 'Lessons to Be Learned from an Experience of Completely Private Financing', *Routes/Roads*, vol. 11, no. 283, 1994, pp. 82–5.
20. Apparently up to ten years.
21. Antonio Estache and Ginés de Rus, eds., *Privatization and Regulation of Transport Infrastructure: Guidelines for Policymakers and Regulators*, World Bank Development Studies (Washington, DC: World Bank, 2000), p. 246.
22. Asian Development Bank, *Developing Best Practices*.
23. World Bank, *World Development Report 1994*.
24. Gómez-Ibáñez and Meyer, *Privatization and Regulation*, and Gordon J. Fielding and Daniel B. Klein, 'How to Franchise Highways', *Journal of Transport Economics and Policy*, vol. 27, no. 2, May 1993, pp. 113–30. For the

experience with privatisation of public transport, see Matthew G. Karlaftis, Jason S. Wasson and Erin S. Steadham, 'Impacts of Privatization on the Performance of Urban Transit Systems', *Transportation Quarterly*, vol. 51, no. 3, summer 1997, pp. 67–79; and Richmond, *The Private Provision of Public Transport*.
25. World Bank, *World Development Report 1994*, Major Projects Association, *Beyond 2000*, p. 265, and Asian Development Bank, *Developing Best Practices*.
26. Major Projects Association, *Beyond 2000*, pp. 262–4.
27. Asian Development Bank, *Developing Best Practices*.
28. Estache and de Rus, eds., *Privatization and Regulation*, p. 239.
29. *Ibid.*, pp. 256–7.
30. Asian Development Bank, *Developing Best Practice*.
31. National Audit Office, *Private Finance Initiative*.
32. Nils Bruzelius: *Köpa väg eller köpa vägtjänster: En analys av nya metoder för finansiering av vägar?* (Lund: Institute of Economic Research, Lund University, 1998). For more on privatisation in Germany and Great Britain, see Nikolaos Zahariadis and Christopher S. Allen, 'Ideas, Networks, and Policy Streams: Privatization in Britain and Germany', *Policy Studies Review*, vol. 14, nos. 1–2, spring–summer 1995, pp. 71–98. Regarding privatisation in Europe and its relations to regulatory change, see Michael Moran and Tony Prosser, eds., *Privatization and Regulatory Change in Europe* (Milton Keynes: Open University Press, 1994).
33. A recent report shows that it is not possible to make use of the capital market and shadow tolls in order to reduce risk costs on account of the uncertainty of future traffic volumes, and that the actual shadow tolling structure used today bears this out. For DBFO projects, the revenues of the contractors – in this case special project companies – are thus not a direct function of the traffic volume; rather the average revenue per vehicle decreases – normally sharply – with the volume of traffic. The relationship between the traffic volume and the annual revenues is thereby weakened significantly, which can be seen as the response of the capital market to the fact it has been asked to do something it is not really capable of. The response has in other words been to design the shadow toll structure in such a way that future traffic is not a primary determining factor of the financial performance of the project. See Bruzelius, *Köpa väg eller köpa vägtjänster*.
34. Danish ministries of Finance and Trade, *Alternativ finansiering av infrastrukturprojekt* (Stockholm Finansdepartementet and Näringsdepartementet, 2000).
35. Bruzelius, *Köpa väg eller köpa vägtjänster*.
36. C. M. Tam, 'Build-Operate-Transfer Model for Infrastructure Developments in Asia: Reasons for Successes and Failures', *International Journal of Project Management*, vol. 17, no. 6, December 1999, pp. 377–82.
37. Major Projects Association, *Beyond 2000*, p. 174. See also Cal Clark, John G. Heilman and Gerald W. Johnson, 'Privatization: Moving Beyond Laissez Faire', *Policy Studies Review*, vol. 14, nos. 3–4, autumn–winter 1995, pp. 395–406.

10 FOUR INSTRUMENTS OF ACCOUNTABILITY

1. See also Barry Bozeman, 'Exploring the Limits of Public and Private Sectors: Sector Boundaries as Maginot Line', *Public Administration Review*, vol. 48, no. 2, 1988, pp. 672–3; Christine Kessides, *Institutional Options for the Provision of Infrastructure*, World Bank Discussion Papers, no. 212 (Washington, DC: World Bank, 1993); Francis J. Leazes Jr, 'Public Accountability: Is It a Private Responsibility?', *Administration and Society*, vol. 29, no. 4, September 1997, pp. 395–41; Robert S. Gilmour and Laura S. Jensen, 'Reinventing Government Accountability: Public Functions, Privatization, and the Meaning of "State Action"', *Public Administration Review*, vol. 58, no. 3, May–June 1998, pp. 247–58.
2. With the exception of the company Dangas under the Danish natural gas company Dong.
3. Danish Ministry of Finance, *Erfaringer med statslige aktieselskaber* (Copenhagen: Finansministeriet (Ministry of Finance), 1993), p. 82.
4. See also Geoffrey Vickers, *The Art of Judgment: A Study of Policy Making* (London: Methuen, 1968), chap. 12; Robert C. Seamans, Jr and Frederick I. Ordway, 'The Apollo Tradition: An Object Lesson for the Management of Large-Scale Technological Endeavors', *Interdisciplinary Science Reviews*, vol. 2, no. 4, 1977, pp. 270–304; Day and Klein, *Accountabilities*; B. Rosen, *Holding Government Bureaucracies Accountable*, 2nd edn (New York: Praeger, 1989); Michael M. Harmon, *Responsibility as Paradox: A Critique of Rational Discourse on Government* (Thousand Oaks, CA: Sage, 1995), chap. 7; Bovens, *The Quest for Responsibility: Accountability and Citizenship in Complex Organisations*.
5. Jean-Michel Fourniau, 'Making the Decision More Transparent: Recent Changes in the Treatment of Major Transport Infrastructure Projects in France', unpublished paper, INRETS-DEST, Arcueil, France, undated.
6. For more on public participation and effective communication, see Richard D. Schwass and Brenda Fowler, 'Public Involvement Throughout the Big Chute Hydroelectric Redevelopment Project', *Impact Assessment*, vol. 11, no. 4, 1993; and Toddi A. Steelman and William Ascher, 'Public Involvement Methods in Natural Resource Policy Making: Advantages, Disadvantages and Trade-Offs', *Policy Sciences*, vol. 31, no. 2, 1997, pp. 71–90.
7. Regarding the stakeholder approach to participation, see John A. Altman and Ed Petkus, Jr, 'Towards a Stakeholder-Based Policy Process: An Application of the Social Marketing Perspective to Environmental Policy Development', *Policy Sciences*, vol. 27, no. 1, 1994, pp. 37–51; John MacArthur, 'Stakeholder Analysis in Project Planning: Origins, Applications and Refinements of the Method', *Project Appraisal*, vol. 12, no. 4, December 1997, pp. 251–65; S. Tuler, 'Development of Mutual Understanding Among Stakeholders in Environmental Policy Disputes', in *Human Ecology: Progress Through Integrative Perspectives*, proceedings of the 7th Society for Human Ecology Meeting (East Lansing, MI: Society for Human Ecology, April 1994), pp. 280–84; Stephen M. Born *et al.*, 'Socioeconomic and Institutional Dimensions of Dam Removals: The Wisconsin Experience', *Environmental Management*, vol. 22, no. 3, 1998, pp. 359–70; Bert Enserink, 'A Quick Scan for

Infrastructure Planning: Screening Alternatives Through Interactive Stakeholder Analysis', *Impact Assessment and Project Appraisal*, vol. 18, no. 1, March 2000, pp. 15–22.

8. Kem Lowry, Peter Adler and Neal Milner, 'Participating the Public: Group Process, Politics, and Planning', *Journal of Planning Education and Research*, vol. 16, no. 3, spring 1997, pp. 177–87; Cheryl Simrell King, Kathryn M. Feltey and Bridget O'Neill Susel, 'The Question of Participation: Toward Authentic Public Participation in Public Administration', *Public Administration Review*, vol. 58, no. 4, July–August 1998, pp. 317–26.

9. Interorganizational Committee on Guidelines and Principles for Social Impact Assessment, *Guidelines and Principles for Social Impact Assessment*, NOAA Technical Memorandum/ NMFS-F/SPO-16 (Washington, DC: US Department of Commerce, 1994); John Sinclair and Alan Diduck, 'Public Education: An Undervalued Component of the Environmental Assessment Public Involvement Process', *Environmental Impact Assessment Review*, vol. 15, no. 3, 1995, pp. 219–40; Liette Vasseur et al., 'Advisory Committee: A Powerful Tool for Helping Decision Makers in Environmental Issues', *Environmental Management*, vol. 21, no. 3, 1997, pp. 359–65.

10. Renn, Webler and Wiedemann, eds., *Fairness and Competence in Citizen Participation: Evaluating Models for Environmental Discourse*. See also Daniel J. Fiorino, 'Citizen Participation and Environmental Risk: A Survey of Institutional Mechanisms', *Science, Technology, and Human Values*, vol. 15, 1990, pp. 226–43; O. Renn, T. Webler, H. Rakel, B. Johnson and P. Dienel, 'A Three-Step Procedure for Public Participation in Decision Making', *Policy Sciences*, vol. 26, 1993, pp. 189–214; S. Tuler, 'Learning Through Participation', *Human Ecology Review*, vol. 5, no. 1, 1998, pp. 58–60; S. Tuler and T. Webler, 'Voices From the Forest: What Participants Expect of a Public Participation Process', *Society and Natural Resources*, vol. 12, 1999, pp. 437–53; T. Webler, 'Organizing Public Participation: A Review of Three Handbooks', *Human Ecology Review*, vol. 3, no. 1, 1997, pp. 245–54.

11. Here and in the following paragraph the quotes are from Funtowicz and Ravetz, 'Three Types of Risk Assessment and the Emergence of Post-normal Science', in Krimsky and Golding, eds., *Social Theories of Risk*, pp. 254, 270–71, 294.

12. See also K. P. Kearns, 'Strategic Management of Accountability', *Public Administration Review*, vol. 54, 1993, pp. 185–92; and C. Pollitt and H. Summa, 'Performance Auditing: Travellers' Tales', in E. Chelimsky and W. Shadish, eds., *Evaluation for the Twenty-First Century: A Resource Book* (Thousand Oaks, CA: Sage, 1997).

13. In the transport sector, performance specifications for investment projects are still rare. Eight road projects procured during the 1990s by way of the DBFO technique in England are said to be based on performance specifications. However, these projects have, in effect, been based on traditional engineering specifications; see National Audit Office, *The Private Finance Initiative: The First Four Design, Build, Finance and Operate Roads Contracts*. Performance-based contracting has been employed in Sweden for a number of road-building projects, albeit in an incomplete version; see, e.g.,

Torsten Grennberg and Bo Grönberg, *Fyra vägbyggen på funktionsentreprenad* (Stockholm: Swedish National Road Administration, 1996). Performance-based contracting is currently being introduced in many countries for routine and periodic maintenance of roads, entailing long-term contracts of up to ten years; see, e.g., G. S. Michel, D. Palsat and C. Cawley, 'Issues Relating to Performance Specifications for Roadways', paper presented at the 1998 Annual Conference of the Transportation Association of Canada; and G. Cabana, G. Liautaud and A. Faiz, 'Areawide Performance-Based Rehabilitation and Maintenance Contracts for Low-Volume Roads', Seventh International Conference on Low-Volume Roads, 1999.

14. Paul Anand *et al.*, 'Performance Auditing in the Public Sector: Approaches and Issues in OECD Member Countries', background paper for PUMA Symposium, OECD, Paris, 6–7 June, 1995.
15. Anthony A. Atkinson, John H. Waterhouse and Robert B. Wells, 'A Stakeholder Approach to Strategic Performance Measurement', *Sloan Management Review*, vol. 38, no. 3, spring 1997, pp. 25–37.
16. See for example Bruzelius: *Köpa väg eller köpa vägtjänster? En analys av nya metoder för finansiering av vägar*. See also Gómez-Ibáñez and Meyer, *Going Private: The International Experience with Transport Privatization*.
17. See also Kessides, *Institutional Options*.
18. See for example Dieter Helm, 'British Utility Regulation: Theory, Practice, and Reform', *Oxford Review of Economic Policy*, vol. 10, no. 3, 1994, pp. 17–39.
19. See SOU (Statens offentliga utredningar), 'Citytunneln i Malmö: Betänkande av utredningen med uppdrag att lämna förslag till genomförande och finansiering av en eventuell Citytunnel i Malmö', SOU 1994:78 (Stockholm, 1994); and Regeringens proposition 1999/2000:66: *Villkoren för järnvägstrafiken på den fasta förbindelsen över Öresund m.m.* (Stockholm: SOU, 16 March 2000).
20. Based on primary data from the Danish State Railways, 26 January 2001.
21. Based on preliminary data.
22. See Chapter 9 and G. Haley, 'Private Finance for Transportation and Infrastructure Projects: A View', *International Journal of Project Management*, vol. 10, no. 2, May 1992, pp. 63–68; and Sheila Farrell, ed., *Financing Transport Infrastructure* (London: PTRC Education and Research Services, 1994).
23. Eichengreen, 'Financing Infrastructure in Developing Countries; Lessons from the Railway Age'.
24. In New Zealand, for example, it has been government policy that state-owned enterprises must borrow on the commercial market and without a guarantee. See also Ian Jones, Hadi Zamani and Rebecca Reehal, *Financing Models for New Transport Infrastructure* (Luxemburg: Office for Official Publications of the European Community, 1996).
25. Danish Parliament, 'Redegørelse af 17/4 85 til Folketinget om en fast forbindelse over Store Bælt', Redegørelse no. R 13, Copenhagen, 1985, 8907; Great Belt Ltd, *Årsberetning 1987* (Copenhagen: Great Belt Ltd, 1988), p. 16.

11 ACCOUNTABLE MEGAPROJECT DECISION MAKING

1. Danish Ministry of Finance, *Erfaringer med statslige aktieselskaber* (Copenhagen: Finansministeriet, 1993), pp. 120, 122.

APPENDIX. RISK AND ACCOUNTABILITY AT WORK: A CASE STUDY

1. The Baltic Sea link study was initiated by the Danish Transport Council in 1994. The study team comprised Bent Flyvbjerg (team leader), Nils Bruzelius and Werner Rothengatter. The team was assisted by Mette K. Skamris and Kim Lynge Nielsen.
2. Danish Parliament, 'Aftale mellem Danmarks regering og Sveriges regering om en fast forbindelse over Øresund', Annex to Danish Parliament, 'Forslag til Lov om anlæg af fast forbindelse over Øresund'.
3. The review was reported in full in Danish Transport Council: *Facts About Fehmarn Belt: Fact-finding Study on a Fixed Link Across Fehmarn Belt*, Report no. 95:02; Danish Transport Council, *Fehmarn Belt: Issues of Accountability. Lessons and Recommendations Regarding Appraisal of a Fixed Link Across Fehmarn Belt*. See also Nils Bruzelius, Bent Flyvbjerg and Werner Rothengatter, 'Big Decisions, Big Risks: Improving Accountability in Mega Projects', *International Review of Administrative Sciences*, vol. 64, no. 3, September 1998, pp. 423–40.
4. The main reports from the feasibility studies are: Danish Ministry of Transport, *Femer Bælt-forbindelsen: Forundersøgelser, Resumérapport* (Copenhagen: Trafikministeriet (Ministry of Transport), 1999); *Femer Bælt-forbindelsen: Økonomiske undersøgelser* (Copenhagen: Trafikministeriet, 1999).
5. Danish Transport Council, *Femer Bælt forundersøgelser: Resumé af konferencen 13. januar 2000* (Copenhagen: Transportrådet (Danish Transport Council), 2000).
6. Fehmarnbelt Development Joint Venture, *Fixed Link Across Fehmarnbelt: Finance and Organisation Enquiry of Commercial Interest*, Report, Executive Summary and Appendices (www.fdjv.com: Author, 2002).
7. Public funds are expected to be used for the connecting links on land.
8. Bundeshaushaltsordnung §7.2; Fehmarnbelt Development Joint Venture, *Fixed Link Across Fehmarnbelt*.
9. This refers to the Danish situation.
10. Contribution from the Danish Ministry of Environment and Energy in answering question from Parliament to the Minister of Transport prompted by proposal to the Parliament to initiate a financial risk analysis of a possible fixed link across Fehmarn Belt, 21 March 1996.
11. Fehmarnbelt Development Joint Venture, *Fixed Link Across Fehmarnbelt*.
12. Answer from the Minister of Transport to question S 2275 from Member of Parliament, 3 August 1995.
13. Answer from the Minister of Transport to question A, B from the Parliament Transport Committee, 1 November 1995.
14. Danish Ministry of Transport, *Femer Bælt-forbindelsen: Økonomiske undersøgelser*; Danish Transport Council, *Femer Bælt forundersøgelser*; Otto Anker

Nielsen, Bent Flyvbjerg and Jens Rørbech, 'Notat om arbejdet med vurdering af trafikanalyser og -prognoser, Femer Bælt', paper presented at the Conference on Fehmarn Belt organised by the Ministry of Transport and the Danish Transport Council, Copenhagen, 13 January 2000.

15. Bjarne Madsen and Chris Jensen-Butler, *De regionaløkonomiske konsekvenser af en fast Femer Bælt forbindelse* (Copenhagen: AKF, 1999).

Bibliography

Note: for alphabetical purposes the names of Danish authors or organisations containing the Danish characters ø and æ are treated as o and ae, the nearest characters to them in appearance in the Roman alphabet.

Aaron, Henry, ed., *Setting National Priorities: Policy for the Nineties*, Washington, DC: Brookings Institution, 1990.

Ahlstrand, Ingemar, *Från särintresse till allmänintresse: Om beslutsunderlagets centrala betydelse med exempel från Öresundsbron och Dennispaketet*, Stockholm: SNS Förlag, 1995.

Alho, Juha M., 'The Accuracy of Environmental Impact Assessments: Skew Prediction Errors', *Ambio*, vol. 21, no. 4, 1992, 322–3.

Al-Khalil, Mohammed I. and Mohammed A. Al-Ghafly, 'Delay in Public Utility Projects in Saudi Arabia', *International Journal of Project Management*, vol. 17, no. 2, 1999, 101–106.

Altman, John A. and Ed Petkus, Jr, 'Towards a Stakeholder-Based Policy Process: An Application of the Social Marketing Perspective to Environmental Policy Development', *Policy Sciences*, vol. 27, no. 1, 1994, 37–51.

Alvares, C. and R. Billorey, *Damning the Narmada: India's Greatest Planned Environmental Disaster*, Penang, Malaysia: Third World Network and Asia-Pacific People's Environment Network, APPEN, 1988.

Anand, Paul and David Shand, *Performance Auditing in the Public Sector: Approaches and Issues in OECD Member Countries*, paper presented at symposium organised by the Public Management Service of the OECD held in Paris, 6–7 June 1995; published Paris: OECD, 1996.

Andersson, Morten and Bo Elling, *Environmental Impact Assessment on Transport Projects in Denmark*, Research Centre for Environmental Assessment, Roskilde University, 1992.

Andersson, Morten, Per Homann Jespeson and Henning Schroll, *Miljøvurdering af transportprojekter: et case studie*, Research Centre for Environmental Assessment, Roskilde University, 1994.

Archer, John and Tracy Dunleavy, 'Rethinking Transportation Financing Policies', *Transportation Quarterly*, vol. 50, no. 3, summer 1996, 131–44.

Arditi, David, Guzin Tarim Akan and San Gurdamar, 'Cost Overruns in Public Projects', *International Journal of Project Management*, vol. 3, no. 4, 1985, 218–25.

Armour, Audrey M., 'Risk Assessment in Environmental Policymaking', *Policy Studies Review*, vol. 12, nos. 3–4, autumn–winter 1993, 178–96.
Arquiaga, M. Carlota, Larry W. Canter and Deborah Imel Nelson, 'Risk Assessment Principles in Environmental Impact Studies', *The Environmental Professional*, vol. 14, no. 3, 1992, 204–19.
Arrow, Kenneth J., 'The Theory of Risk-Bearing: Small and Great Risks', *Journal of Risk and Uncertainty*, vol. 12, nos. 2–3, 1996, 103–11.
Arrow, Kenneth J., and Robert C. Lind, 'Uncertainty and the Evaluation of Public Investment Decisions', *American Economic Review*, vol. 60, 1970, 364–78.
A/S Storebæltsforbindelsen, *Øst-vest trafikmodellen: Prognoser for trafikken mellem Øst- og Vestdanmark*, Copenhagen: Great Belt Ltd, February 1991.
Øst-vest trafikmodellen: Prognoser for trafikken over Storebæltsbroen og de konkurrerende færgeruter, Copenhagen: Great Belt Ltd, August 1994.
Aschauer, D. A., 'Is Public Expenditure Productive?', *Journal of Monetary Economics*, vol. 23, 1989, 177–200.
Public Infrastructure Investment: A Bridge to Productivity Growth?, Public Policy Brief 4, New York: Bard College, Jerome Levy Economic Institute, Annandale-on-Hudson, 1993.
Ascher, Kate, *The Politics of Privatization: Contracting Out Public Services*, New York: St Martin's Press, 1987.
Asian Development Bank, *Developing Best Practices for Promoting Private Sector Investment in Infrastructure: Roads*, Manila, Philippines, 2000.
Atkinson, Anthony A., John H. Waterhouse and Robert B. Wells, 'A Stakeholder Approach to Strategic Performance Measurement', *Sloan Management Review*, vol. 38, no. 3, spring 1997, 25–37.
Bailey, Elizabeth E. and Janet Rothenberg Pack, eds., *The Political Economy of Privatization and Deregulation*, Aldershot: Edward Elgar, 1995.
Bauman, Zygmunt, *Globalization: The Human Consequences*, Cambridge: Polity Press, 1998.
'Time and Class: New Dimensions of Stratification', *Sociologisk Rapportserie*, no. 7, Department of Sociology, University of Copenhagen, 1998.
Beattie, Robert B., 'Everything You Already Know About EIA (But Don't Often Admit)', *Environmental Impact Assessment Review*, vol. 15, no. 2, March 1995, 109–14.
Beck, Ulrich, *Risk Society: Towards a New Modernity*, Thousand Oaks, CA: Sage, 1992.
Beidleman, Carl R., Donna Fletcher and David Veshosky, 'On Allocating Risk: The Essence of Project Finance', *Sloan Management Review*, vol. 31, no. 3, spring 1990, 47–55.
Bein, Peter and Mike Kawczynski, 'Environmental Accounting Applied to Greater Vancouver Transportation System Planning', paper presented to the 76th Annual Meeting of the Transportation Research Board, Washington, DC: Transportation Research Board, January 1997.
Biehl, Dieter, *The Contribution of Infrastructure to Regional Development*, Final Report, Infrastructure Study Group, Luxembourg: Commission of the European Communities, 1986.

Bignell, Victor and Joyce Fortune, *Understanding Systems Failures*, Manchester University Press, 1984.

Bird, Ashley and Riki Therivel, 'Post-Auditing of Environmental Impact Statements Using Data Held in Public Registers of Environmental Information', *Project Appraisal*, vol. 11, no. 2, June 1996, 105–16.

Bisset, Ronald, 'Post Development Audits to Investigate the Accuracy of Environmental Impact Predictions', *Zeitschrift für Umweltpolitik*, vol. 7, no. 4, 1984, 463–84.

Blackshaw, Philip, John Flora and Richard Scurfield, 'Motorways by BOT: Political Dogma or Economic Rationality?', in *Financing Transport Infrastructure*, London: PTRC Publications, 1992, 1–13.

Blake, Coleman, David Cox and Willard Fraize, *Analysis of Projected Vs. Actual Costs for Nuclear and Coal-Fired Power Plants*, Prepared for the United States Energy Research and Development Administration, McLean, VI: Mitre Corporation, 1976.

Blanc, André, Christian Brossier, Christian Bernardini and Michel Gerard, *Rapport de la Mission sur la Projet de TGV-Est Européen*, Paris: Inspection Générale des Finances et Conseil Général des Pont et Chausées, July 1996.

Bohman, James, *Public Deliberation: Pluralism, Complexity, and Democracy*, Cambridge, MA: MIT Press, 1996.

Boholm, Åsa and Ragnar Löfstedt, 'Issues of Risk, Trust and Knowledge: The Hallandsås Tunnel Case', *Ambio*, vol. 28, no. 6, September 1999, 556–61.

Bond, Gary and Laurence Carter, 'Financing Private Infrastructure Projects; Emerging Trends from IFC's Experience', *IFC Discussion Paper*, No. 23, Washington, DC: IFC 1994.

Boothroyd, Peter, Nancy Knight, Margaret Eberle, June Kawaguchi and Christiane Gaguon, 'The Need for Retrospective Impact Assessment: The Megaprojects Example', *Impact Assessment*, vol. 13, no. 3, September 1995, 253–71.

Born, Stephen M., Kenneth D. Genskow, Timothy L. Filbert, Nuria Hernaudez-Mora, Matthew L. Keefer and Kimberly A. White, 'Socioeconomic and Institutional Dimensions of Dam Removals: The Wisconsin Experience', *Environmental Management*, vol. 22, no. 3, 1998, 359–70.

Bovens, Mark, *The Quest for Responsibility: Accountability and Citizenship in Complex Organisations*, Cambridge University Press, 1998.

Bovens, Mark and Paul 't Hart, *Understanding Policy Fiascoes*, New Brunswick, NJ: Transaction Publishers, 1996.

Boyce, Tim, *Commercial Risk Management: How to Identify, Mitigate and Avoid the Principal Risks in Any Project*, London: Thorogood, 1995.

Bozeman, Barry, 'Exploring the Limits of Public and Private Sectors: Sector Boundaries as Maginot Line', *Public Administration Review*, vol. 48, no. 2, 1988, 672–3.

Bradshaw, B. and R. Vickerman, eds., *The Channel Tunnel: Transport Studies*, In Focus, Folkestone: The Channel Tunnel Group Ltd, 1993.

Brundtland Commission, *Our Common Future*, Oxford University Press, 1987.

Bruzelius, Nils, *Köpa väg eller köpa vägtjänster? En analys av nya metoder för finansiering av vägar*, Institute of Economic Research, Lund University, 1998.

Bruzelius, Nils, Bent Flyvbjerg and Werner Rothengatter, 'Big Decisions, Big Risks: Improving Accountability in Mega Projects', *International Review of Administrative Sciences*, vol. 64, no. 3, September 1998, 423–40.
Buckley, Ralf C., 'Environmental Audit: Review and Guidelines', *Environment and Planning Law Journal*, vol. 7, no. 2, June 1990, 127–41.
'Auditing the Precision and Accuracy of Environmental Impact Predictions in Australia', *Environmental Monitoring and Assessment*, vol. 18, July 1991, 1–23.
'How Accurate are Environmental Impact Predictions?', *Ambio*, vol. 20, nos. 3–4, 1991, 161–2.
'Response to Comment by J. M. Alho', *Ambio*, vol. 21, no. 4, 1992, 323–4.
BVU, Hensch and Boesefeldt and PLANCO, *Analyse des BVWP-Verfahrens in Methodik und Ablauf*, Bonn: Federal Ministry for Transport.
Byron, Helen J., Joanna R. Treweek, William R. Sheate and Stewart Thompson, 'Road Developments in the UK: An Analysis of Ecological Assessment in Environmental Impact Statements Produced between 1993 and 1997', *Journal of Environmental Planning and Management*, vol. 43, no. 1, January 2000, 71–97.
Cabana, G., G. Liautaud and A. Faiz, 'Areawide Performance-Based Rehabilitation and Maintenance Contracts for Low-Volume Roads', Seventh International Conference on Low-Volume Roads, 1999.
Cairncross, Frances, *The Death of Distance: How the Communications Revolution Will Change Our Lives*, Boston, MA: Harvard Business School Press, 1997.
Canaday, Henry T., *Construction Cost Overruns in Electric Utilities: Some Trends and Implications*, Occasional Paper no. 3, Columbus: National Regulatory Research Institute, Ohio State University, November 1980.
Canter, Larry W., 'Impact Prediction Auditing', *The Environmental Professional*, vol. 7, no. 3, 1985, 255–64.
'The Role of Environmental Monitoring in Responsible Project Management', *The Environmental Professional*, vol. 15, no. 1, 1993, 76–87.
Carnemark, Curt, Jaime Biderman and David Bovet, 'The Economic Analysis of Rural Road Projects', *World Bank Staff Working Papers*, no. 214, Washington, DC: World Bank, 1984.
Cervero, Robert and John Landis, 'Assessing the Impacts of Urban Rail Transit on Local Real Estate Markets Using Quasi-Experimental Comparisons', *Transportation Research*, Part A, vol. 27A, no. 1, 1993, 13–22.
Channel Tunnel Group, *The Channel Tunnel Project: Environmental Effects in the UK*, Folkestone: Channel Tunnel Group Ltd, 1985.
Charest, Paul, 'Aboriginal Alternatives to Megaprojects and Their Environmental and Social Impacts', *Impact Assessment*, vol. 13, no. 4, 1995, 371–86.
Charpentier, G., 'Lessons to Be Learned from an Experience of Completely Private Financing', *Routes/Roads*, vol. 11, no. 283, 1994, 82–5.
Christoffersen, Mads, Bent Flyvbjerg and Jørgen Lindgaard Pedersen, 'The Lack of Technology Assessment in Relation to Big Infrastructural Decisions', in *Technology and Democracy: The Use and Impact of Technology Assessment in Europe. Proceedings from the 3rd European Congress on Technology Assessment*, vol. I, Copenhagen: n.p., 4–7 November 1992, 54–75.

Clark, Cal, John G. Heilman and Gerald W. Johnson, 'Privatization: Moving Beyond Laissez Faire', *Policy Studies Review*, vol. 14, nos. 3-4, autumn-winter 1995, 395-406.

Cnotka, Martin Hayo Hermann, Chris Jensen-Bulter, Erik Christiansen, Bjarne Madsen, et al., *Sociøøkonomisk analyse af Storstrøms Amt og Kreis Ostholstein*, Copenhagen: AKF Forlaget, February 1994.

Collingridge, David, *The Management of Scale: Big Organizations, Big Decisions, Big Mistakes*, London: Routledge, 1992.

'Undemocratic Technology: Big Organization, Big Governments, Big Projects and Big Mistakes', paper presented at Technological Strategies for the New Europe, International Seminar, UNESCO, undated.

Commission of the European Union, *Evaluation of the Effects of the Cross-Channel Fixed Link on Traffic Flows*, COST 312, Brussels: Commission of the European Union, 1993.

Cooper, Dale and Chris Chapman, *Risk Analysis for Large Projects: Models, Methods, and Cases*, New York: John Wiley & Sons, 1987.

Coote, Anna, 'Risk and Public Policy: Towards a High-Trust Democracy', in Jane Franklin, ed., *The Politics of Risk Society*, Cambridge: Polity Press, 1998, 124-31.

Couillard, Jean, 'The Role of Project Risk in Determining Project Management Approach', *Project Management Journal*, vol. 26, no. 4, 1995, 3-15.

Council of Standards Australia and Council of Standards New Zealand, *Risk Management*, AS/NZS 4360:1995, Homebush: Standards Australia, and Wellington: Standards New Zealand, 1995.

Culhane, Paul J., 'The Precision and Accuracy of U.S. Environmental Impact Statements', *Environmental Monitoring and Assessment*, vol. 8, no. 3, 1987, 217-38.

Cummings, Joe, *Thailand*, Melbourne: Lonely Planet Publications, 1999.

Danish Auditor-General, *Beretning til statsrevisorerne om udviklingen i de økonomiske overslag vedrørende Øresundsforbindelsen*, RB 2001/94, Copenhagen: Rigsrevisionen, November 1994.

Danish Ministries of Finance and Trade, *Alternative finansiering av infrastrukturprojekt*, Stockholm: Finansdepartementet and Näringsdepartementet, 2000.

Danish Ministry of Finance, *Erfaringer med statslige aktieselskaber*, Copenhagen: Finansministeriet, 1993.

Danish Ministry of Transport, *Trafik 2005: Problemstillinger, mål og strategier*, Copenhagen: Trafikministeriet, 1993.

Undersøgelser vedrørende Femer Bælt-forbindelsen: Fase 1 af kyst til kyst undersøgelserne, Copenhagen: Trafikministeriet, 1996.

Femer Bælt-forbindelsen: Forundersøgelser, Resumérapport, Copenhagen: Trafikministeriet, 1999.

Femer Bælt-forbindelsen: Økonomiske undersøgelser, Copenhagen: Trafikministeriet, 1999.

Danish Ministry of Transport and German Federal Ministry for Transport, 'Project Description of Geological, Technical and Environmental Investigations Concerning the Coast-to-Coast Section of a Fixed Link Across Fehmarn Belt', Copenhagen and Bonn: Trafikministeriet and Bundesministerium für Verkehr, 9 December 1994.

Bibliography 185

Femer Belt Crossing Technical and Environmental Feasibility Study Copenhagen: and Bonn March 1993.
Phase A, Doc. no. A1-1, *Technical Solutions*.
Phase A, Doc. no. A2-1, *Environmental Site Conditions*.
Phase A, Doc. no. A2-2, *Technical Site Conditions*.
Phase A, Doc. no. A4-1, *Preliminary Design Basis*.
Phase A, Doc. no. A5-1, *Preliminary Costing Basis*.
Phase A, Doc. no. A6-1, *Planning of Further Environmental Studies*.
Phase A, Doc. no. A6-2, *Planning of Additional Technical Site Investigations*.
Phase A, Doc. no. A6-3, *Planning of Development of Technical Concepts*.
Phase A, Doc. no. A7-1, *Summary Report*.
'Project Description of Geological, Technical and Environmental Investigations Concerning the Coast-to-Coast Section of a Fixed Link Across Fehman Belt', Copenhagen: Danish Ministry of Transport, and Bonn: German Federal Ministry for Transport, 9 December 1994.
Danish Parliament, 'Redegørelse af 17/4 85 til Folketinget om en fast forbindelse over Store Bælt', Redegørelse no. R 13, Copenhagen: Folketinget, 1985.
'Bemærkninger til Forslag til Lov om anlæg af fast forbindelse over Øresund', Lovforslag no. L 178, Folketinget 1990–91, 2. samling, proposed 2 May 1991, p. 10.
'Redegørelse af 6/12 93 om anlæg af fast forbindelse over Øresund', Fortryk af Folketingets forhandlinger, Copenhagen: Folketinget, 7 December 1993.
Danish Parliamentary Auditor's Committee, *Beretning om Storebæltsforbindelsens økonomi*, Beretning 4/97, Copenhagen: Statsrevisoratet (Danish Auditor's Office), 1998.
Danish Parliamentary Committee on Public Works, 'Udvalgets spøgsmål til ministeren for offentlige arbejder og dennes besvarelse heraf', annex to 'Betænkning over Forslag til Lov om anlæg af fast forbindelse over Storebælt', report from the Parliamentary Committee on Public Works, 19 May 1987, p. 21.
Danish State Railways, *Årsrapport 2001* (www.dsb.dk: Author, 2002).
Danish Transport Council, *Facts About Fehmarn Belt: Fact-finding Study on a Fixed Link Across Fehmarn Belt*, Report no. 95:02, Copenhagen: Danish Transport Council, 1995.
Fehmarn Belt: Issues of Accountability: Lessons and Recommendations Regarding Appraisal of a Fixed Link Across Fehmarn Belt, Report no. 95:03, Copenhagen: Danish Transport Council, 1995.
'Regionaløkonomiske effekter af investeringer i trafikinfrastruktur: Notat til Trafikministeren', in Transportrådet, *Fire baggrundsnotater til trafik 2005*, Notat no. 93:09, Copenhagen: Transportrådet, December 1993, 1–18.
Femer Bælt forundersøgelser: Resumé af konferencen 13. januar 2000, Copenhagen: Transportrådet, 2000.
Davidson, Frank P. and Jean-Claude Huot, 'Management Trends for Major Projects', *Project Appraisal*, vol. 4, no. 3, September 1989, 133–42.
Day, P. and R. Klein, *Accountabilities*, London: Tavistock, 1987.
Den svenska Öresundsdelegationens Plan- och Miljögrupp, *Öresundsförbindelser; Planerings- och miljöfrågor*, DsK 1978:6, Stockholm: Den svenska Öresundsdelegation, 1978.

Department of Energy Study Group, *North Sea Costs Escalation Study*, Energy Paper no. 8, London: DDE, 31 December 1975.

Department of the Environment, Transport and the Regions, *Transport and the Economy*, London: Standing Advisory Committee on Trunk Road Assessment, 1999.

Department of Finance, Australia, 'The Choice of Discount Rate for Evaluating Public Sector Investment Projects: A Discussion Paper', Canberra: Department of Finance, November 1987.

Dipper, Ben, Carys Jones and Christopher Wood, 'Monitoring and Post-auditing in Environmental Impact Assessment: A Review', *Journal of Environmental Planning and Management*, vol. 41, no. 6, November 1998, 731–47.

Dlakwa, M. M. and M. F. Culpin, 'Reasons for Overrun in Public Sector Construction Projects in Nigeria', *International Journal of Project Management*, vol. 8, no. 4, 1990, 237–40.

Dodgson, J. S., 'Motorway Investment, Industrial Transport Costs, and Subregional Growth: A Case Study of the M62', *Regional Studies*, vol. 8, 1973, 75–91.

Doherty, Brian, 'Paving the Way: The Rise of Direct Action Against Road-Building and the Changing Character of British Environmentalism', *Political Studies*, vol. 47, no. 2, June 1999, 275–91.

Dryzek, John S., *Deliberative Democracy and Beyond: Liberals, Critics, Contestations*, Oxford: Oxford University Press, 2000.

Duer, Henrik, 'Øst–Vest trafikkens energiforbrug næsten uændret', *Transportrådets Nyhedsbrev*, no. 3, September 1999, 9–10.

'Energiforbrug i Øst–Vest trafikken', unpublished paper, Lyngby: COWI, September 1999.

Economist, The, 19 August 1985.

29 April 1989, p. 73.

'Under water, Over Budget', 7 October 1989, pp. 37–8.

30 April 1994, p. 73.

24 May 1997, pp. 67–8.

31 January 1998, p. 74.

Eichengreen, Barry, 'Financing Infrastructure in Developing Countries: Lessons from the Railway Age', *Policy Research Working Paper*, no. 1379, Washington, DC: World Bank, November 1994.

Elster, Jon, ed., *Deliberative Democracy*, Cambridge University Press, 1998.

Enquete-Kommission des Deutschen Bundestages, 'Vorsorge zum Schutz der Erdatmosphäre', Bundestagsdrucksache 12/8300, Bonn: Enquete-Kommission des Deutschen Bundestages, 1994.

Enserink, Bert, 'A Quick Scan for Infrastructure Planning: Screening Alternatives Through Interactive Stakeholder Analysis', *Impact Assessment and Project Appraisal*, vol. 18, no. 1, March 2000, 15–22.

Estache, Antonio and Ginés de Rus, eds., *Privatization and Regulation of Transport Infrastructure: Guidelines for Policymakers and Regulators*, World Bank Development Studies, Washington, DC: World Bank, 2000.

EU Commission, *Guidelines for the Construction of a Transeuropean Transport Network*, EU Bulletin L228, Brussels: EU Commission, 1996.

Bibliography 187

Falkemark, Gunnar, *Öresundsbron: Hur de avgörande besluten togs*, Report, Stockholm: Naturskyddsföreningen, 1992.
  Politik, lobbyism och manipulation: Svensk trafikpolitik i verkligheten, Nora: Nya Doxa, 1999.
Farrell, Sheila, ed., *Financing Transport Infrastructure*, London: PTRC Education and Research Services, 1994.
Fayard, Alain, 'Overview of the Scope and Limitations of Public–Private Partnerships', Seminar of Public–Private Partnerships (PPPs) in Transport Infrastructure Financing, European Conference of Ministers of Transport, Paris, 1999.
Fearnside, Philip M., 'The Canadian Feasibility Study of the Three Gorges Dam Proposed for China's Yangzi River: A Grave Embarrassment to the Impact Assessment Profession', *Impact Assessment*, vol. 12, no. 1, spring 1994, 21–57.
Fehmarnbelt Development Joint Venture, *Fixed Link Across Fehmarnbelt: Finance and Organisation Enquiry of Commercial Interest*, Report, Executive Summary and Appendices (www.fdjv.com: Author, 2002).
Feldman, Elliot J., 'Patterns of Failure in Government Megaprojects: Economics, Politics, and Participation in Industrial Democracies', in Samuel P. Huntington and Joseph S. Nye, Jr, eds., *Global Dilemmas*, Cambridge, MA: Harvard University, Center for International Affairs, and Lanham, MD: University Press of America, 1985, 138–58.
Fielding, Gordon J. and Daniel B. Klein, 'How to Franchise Highways', *Journal of Transport Economics and Policy*, vol. 27, no. 2, May 1993, 113–30.
Fiorino, Daniel J., 'Citizen Participation and Environmental Risk: A Survey of Institutional Mechanisms', *Science, Technology, and Human Values*, vol. 15, 1990, 226–43.
Flyvbjerg, Bent, 'Implementation and the Choice of Evaluation Methods', *Transport Policy and Decision Making*, vol. 2, no. 3, 1984, 291–314.
  'The Open Format and Citizen Participation in Transportation Planning', in Transportation Research Board, *Social and Technological Issues in Transportation Planning*, Washington, DC: National Research Council, 1984, 15–22.
  'The Dark Side of Planning: Rationality and *Realrationalität*', in Seymour Mandelbaum, Luigi Mazza and Robert Burchell, eds., *Explorations in Planning Theory*, New Brunswick, NJ: Center for Urban Policy Research Press, 1996, 383–94.
  'Habermas and Foucault: Thinkers for Civil Society?', *British Journal of Sociology*, vol. 49, no. 2, June 1998, 208–33.
  Rationality and Power: Democracy in Practice, Chicago: University of Chicago Press, 1998.
  Making Social Science Matter: Why Social Inquiry Fails and How It Can Succeed Again, Cambridge University Press, 2001.
  'Economic Risk in Public Works Projects: The Case of Urban Rail', paper, forthcoming.
Flyvbjerg, Bent, Uffe Jacobsen, Kai Lemberg and Poul Ove Pedersen, 'Transport, Communications and Mobility in Denmark', in Peter NijKamp, Shalom

Reichman and Michael Wegener, eds., *Euromobile: Transport, Communications and Mobility in Europe*, Strasbourg: European Science Foundation, and Aldershot: Avebury-Gower, 1990, 149–57.

Flyvbjerg, Bent, Kim Lynge Nielsen and Mette K. Skamris, 'Research on Major Transport Projects: Economy, Environment and Institutions', paper presented at Colloquium on Integrated Strategic Infrastructure Networks in Europe, 2–3 March 1995, Lausanne, Switzerland.

Flyvberg, Bent, and Mette K. Skamris Holm, 'How Accurate are Demand Forecasts in Transport Infrastructure Projects', paper, forthcoming.

Flyvbjerg, Bent, Mette K. Skamris Holm and Søren L. Buhl, 'Underestimating Costs in Public Works Projects: Error or Lie?', *Journal of the American Planning Association*, vol. 68, no. 3, summer 2002, pp. 279–95.

Fouracre, P. R., R. J. Allport and J. M. Thomson, *The Performance and Impact of Rail Mass Transit in Developing Countries*, TRRL Research Report 278. Crowthorne: Transport and Road Research Laboratory, 1990.

Fourniau, Jean-Michel, 'Making the Decision More Transparent: Recent Changes in the Treatment of Major Transport Infrastructure Projects in France', unpublished paper, INRETS-DEST, Arcueil, France, undated.

Franklin, Jane, ed., *The Politics of Risk Society*, Cambridge: Polity Press, 1998.

Fraser, R. M., 'Compensation for Extra Preliminary and General (P & G) Costs Arising from Delays, Variations and Disruptions: The Palmiet Pumped Storage Scheme', *Tunnelling and Underground Space Technology*, vol. 5, no. 3, 1990, 205–16.

Funtowicz, Silvio O. and Jerome R. Ravetz, 'Three Types of Risk Assessment and the Emergence of Post-normal Science', in Sheldon Krimsky and Dominic Golding, eds., *Social Theories of Risk*, Westport, CT: Praeger, 1992, 251–73.

Garett, Mark and Martin Wachs, *Transportation Planning on Trial: The Clean Air Act and Travel Forecasting*, Thousand Oaks, CA: Sage, 1996.

Gaudry, M., B. Mandel and W. Rothengatter, 'Linear and Nonlinear Logit Models', *Transportation Research*, part B, vol. 28, no. 6, 1994, 91–102.

German Federal Ministry for Transport, *Bundesverkehrswegeplan 1992*, Bonn: Bundesministerium für Verkehr, 1992.

'Gesamtwirtschaftliche Bewertung von Verkehrsinfrastrukturinvestitionen für den Bundesverkehrswegeplan', *Schriftenreihe des Bundesministeriums für Verkehr*, Heft 72, Bonn: Bundesministerium für Verkehr, 1992.

Giddens, Anthony, *The Consequences of Modernity*, Stanford, CA: Stanford University Press, 1990.

Gilmour, Robert S. and Laura S. Jensen, 'Reinventing Government Accountability: Public Functions, Privatization, and the Meaning of "State Action"', *Public Administration Review*, vol. 58, no. 3, May–June 1998, 247–58.

Gilpin, Alan, *Environmental Impact Assessment (EIA): Cutting Edge for the Twenty-First Century*, Cambridge University Press, 1995.

Glasson, John, 'Life After the Decision: The Importance of Monitoring in EIA', *Built Environment*, vol. 20, no. 4, 1994, 309–320.

Glasson, John, Riki Therival, Joe Weston, Elizabeth Wilson and Richard Frost, 'EIA – Learning from Experience: Changes in the Quality of Environmental

Impact Statements for UK Planning Projects', *Journal of Environmental Planning and Management*, vol. 40, no. 4, 1997, 451–64.
Goetz, Andrew R. and Joseph S. Szyliowicz, 'Revisiting Transportation Planning and Decision Making Theory: The Case of Denver International Airport', *Transportation Research*, part A, vol. 31, no. 4, 263–80.
Gómez-Ibáñez, José A. and John R. Meyer, *Going Private: The International Experience with Transport Privatization*, Washington, DC: Brookings Institution, 1993.
Gramlich, Edward M., 'Infrastructure Investment: A Review Essay', *Journal of Economic Literature*, vol. 32, September 1994, 1176–96.
Grayson, Lesley, *Environmental Auditing: A Guide to Best Practice in the UK and Europe*, Letchworth, UK: Technical Communications in association with the British Library and Information Service, 1992.
Grayson, Lesley, Helen Woolston and Joseph Tanega, *Business and Environmental Accountability: An Overview and Guide to the Literature*, Letchworth, UK: Technical Communications in association with the British Library, 1994.
Great Belt Ltd, *Årsberetning 1987*, Copenhagen: Great Belt Ltd, 1988.
Gregory, R., R. Keeney and D. V. Winterfeldt, 'Adapting the Environmental Impact Statement Process to Inform Decisionmakers', *Journal of Policy Analysis and Management*, vol. 11, no. 1, 1992, 58–75.
Grennberg, Torsten and Bo Grönberg, *Fyra vägbyggen på funktionsentreprenad*, Stockholm: Swedish National Road Administration, 1996.
Gundersen, Adolf G., *The Environmental Promise of Democratic Deliberation*, Madison, WI: University of Wisconsin Press, 1995.
Gunningham, Neil, 'Environmental Auditing: Who Audits the Auditors?', *Environmental and Planning Law Journal*, vol. 10, no. 4, 1993, 229–238.
Haley, G., 'Private Finance for Transportation and Infrastructure Projects: A View', *International Journal of Project Management*, vol. 10, no. 2, May 1992, 63–8.
Hall, Peter, *Great Planning Disasters*, Harmondsworth: Penguin, 1980.
'Great Planning Disasters Revisited', paper, Bartlett School, London, undated.
Hall, Peter, and Rosemary C. R. Taylor, 'Political Science and the Three New Institutionalisms', *Political Studies*, vol. 44, no. 5, 1996, 936–57.
Hanke, Steve H., ed., *Prospects for Privatization*, New York: Academy of Political Science, 1987.
Harmon, Michael M., *Responsibility as Paradox: A Critique of Rational Discourse on Government*, Thousand Oaks, CA: Sage, 1995.
Healey, J. M., 'Errors in Project Cost Estimates', *Indian Economic Journal*, vol. 12, no. 1, July–September 1964, 44–52.
Helm, Dieter, 'British Utility Regulation: Theory, Practice, and Reform', *Oxford Review of Economic Policy*, vol. 10, no. 3, 1994, 17–39.
Henderson, P. D., 'Two British Errors: Their Probable Size and Some Possible Lessons', *Oxford Economic Papers*, vol. 29, no. 2, July 1977, 159–205.
Hertz, David B. and Howard Thomas, *Risk Analysis and Its Applications*, New York: John Wiley & Sons Ltd, 1984.

Hervik, Arild and Svein Bråthen, *Økonomiske analyser av nye bru/tunnel-prosjekter: Fokus på markedsanalyser og nytte/kostnadsanalyser*, Arbeidsrapport no. M 9003, Molde: Forskningscenter Molde, 1990.

Hjalager, Anne-Mette, *Transportinfrastruktur og regional udvikling: Danske undersøgelser*, Transportrådet, Notat no. 93:07, Copenhagen: Danish Transport Council, December 1993.

Hufschmidt, Maynard M. and Jacques Gerin, 'Systematic Errors in Cost Estimates for Public Investment Projects', in Julius Margolis, ed., *The Analysis of Public Output*, New York: Columbia University Press, 1970, 267–315.

Huntington, S. P. and J. S. Nye, Jr, *Global Dilemmas*, Cambridge, MA: Harvard University Press, 1985.

Hurdle, David, 'Does Transport Investment Stimulate Economic Activity?', *The Planner*, vol. 78, no. 9, May 1992, 7–9.

Huszar, Paul C., 'Overestimated Benefits and Underestimated Costs: The Case of the Paraguay-Paraná Navigation Study', *Impact Assessment and Project Appraisal*, vol. 16, no. 4, December 1998, 295–304.

Interorganizational Committee on Guidelines and Principles for Social Impact Assessment, *Guidelines and Principles for Social Impact Assessment*, NOAA Technical Memorandum/NMFS-F/SPO-16, Washington, DC: US Department of Commerce, 1994.

Israel, Arturo, *Issues for Infrastructure Management in the 1990's*, World Bank Discussion Papers, no. 171, Washington, DC: World Bank, 1992.

Jaeger, Carlo, Ortwin Renn, Eugene A. Rosa and Thomas Webler, *Risk, Uncertainty and Rational Action*, London: Earthscan, 2001.

Jones, Ian, Hadi Zamani and Rebecca Reehal, *Financing Models for New Transport Infrastructure*, Luxembourg: Office for Official Publications of the European Community, 1996.

Kain, John F., 'Choosing the Wrong Technology: Or How to Spend Billions and Reduce Transit Use', *Journal of Advanced Transportation*, vol. 21, no. 3, winter 1988, 197–213.

'Deception in Dallas: Strategic Misrepresentation in Rail Transit Promotion and Evaluation', *Journal of the American Planning Association*, vol. 56, no. 2, spring 1990, 184–96.

Kamerman, Sheila B. and Alfred J. Kahn, eds., *Privatization and the Welfare State*, Princeton, NJ: Princeton University Press, 1989.

Karl, Helmut, 'Better Environmental Future in Europe Through Environmental Auditing?', *Environmental Management*, vol. 18, no. 4, 1994, 617–21.

Karlaftis, Matthew G., Jason S. Wasson and Erin S. Steadham, 'Impacts of Privatization on the Performance of Urban Transit Systems', *Transportation Quarterly*, vol. 51, no. 3, summer 1997, 67–79.

Kearns, K. P., 'Strategic Management of Accountability', *Public Administration Review*, vol. 54, 1993, 185–92.

Keller, Reiner and Angelika Poferl, 'Habermas Fightin' Waste: Problems of Alternative Dispute Resolution in the Risk Society', *Journal of Environmental Policy and Planning*, vol. 2, no. 1, 2000, 55–67.

Kessides, Christine, *Institutional Options for the Provision of Infrastructure*, World Bank Discussion Papers, no. 212, 1993, Washington, DC: World Bank, 1993.

Khan, Mushtaq H. and K. S. Jomo, eds., *Rents, Rent-Seeking and Economic Development: Theory and Evidence in Asia*, Cambridge University Press, 2000.

Kharbanda, O. P. and E. A. Stallworthy, *How to Learn from Project Disasters: True-Life Stories with a Moral for Management*, Aldershot: Gower, 1983.

King, Cheryl Simrell, Kathryn M. Feltey and Bridget O'Neill Susel, 'The Question of Participation: Toward Authentic Public Participation in Public Administration', *Public Administration Review*, vol. 58, no. 4, July–August 1998, 317–26.

Koschatzky, Knut, 'A River is a River: Cross-Border Networking Between Baden and Alsace', *European Planning Studies*, vol. 8, no. 4, August 2000, 429–49.

Krieger, Martin H., ed., 'Big Decisions, Big Projects, Big Plans', book manuscript, Los Angeles: University of Southern California, School of Urban Planning and Development, undated.

Krimsky, Sheldon and Dominic Golding, eds., *Social Theories of Risk*, Westport, CT: Praeger, 1992.

Kulash, Damian J., 'Economic Returns from Transportation Investment', *Transportation Quarterly*, vol. 51, no. 3, summer 1997, 8–19.

Lawrence, David P., 'Quality and Effectiveness of Environmental Impact Assessments: Lessons and Insights from Ten Assessments in Canada', *Project Appraisal*, vol. 12, no. 4, December 1997, 219–32.

Leavitt, Dan, Sean Ennis and Pat McGovern, *The Cost Escalation of Rail Projects: Using Previous Experience to Re-evaluate the CalSpeed Estimates*, Working Paper 567, Berkeley: Institute of Urban and Regional Development, University of California, 1993.

Leazes, Francis J., Jr, 'Public Accountability: Is It a Private Responsibility?', *Administration and Society*, vol. 29, no. 4, September 1997, 395–41.

Leitch, Gordon and J. Ernest Tanner, 'Professional Economic Forecasts: Are They Worth Their Costs?', *Journal of Forecasting*, vol. 14, no. 2, 1995, 143–57.

Leu, Wen-Shyan, W. P. Williams and A. W. Bark, 'Evaluation of Environmental Impact Assessment in Three Southeast Asian Nations', *Project Appraisal*, vol. 12, no. 2, June 1997, 89–100.

Lewis, David, 'The Future of Forecasting: Risk Analysis As a Philosophy of Transportation Planning', *TR News*, no. 177, March 1995, 3–9.

Lind, Robert C., ed., *Discounting for Time and Risk in Energy Policy*, Baltimore: Johns Hopkins University Press, 1982.

Lipton, Joshua *et al.*, 'A Paradigm for Ecological Risk Assessment', *Environmental Management*, vol. 17, no. 1, 1993, 1–5.

Lockwood, Stephen, Ravindra Verma and Michael Schneider, 'Public–Private Partnerships in Toll Road Development: An Overview of Global Practices', *Transportation Quarterly*, vol. 54, no. 2, spring 2000, 77–91.

Lowndes, Vivien, 'Varieties of New Institutionalism: A Critical Appraisal', *Public Administration*, vol. 74, summer 1996, 181–97.

Lowry, Kem, Peter Adler and Neal Milner, 'Participating the Public: Group Process, Politics, and Planning', *Journal of Planning Education and Research*, vol. 16, no. 3, spring 1997, 177–87.

Luberoff, David and Alan Altshuler, *Megaproject: A Political History of Boston's Multibillion Dollar Central Artery/Third Harbor Tunnel Project*, Cambridge, MA: Taubman Center for State and Local Government, Kennedy School of Government, Harvard University 1996.

Luery, Andrea D., Luis Vega and Jorge Gastelumendi de Rossi, *Sabotage in Santa Valley: The Environmental Implications of Water Mismanagement in a Large-Scale Irrigation Project in Peru*, Norwalk, CT: Technoserve, 1991.

Luk, Shiu-hung and Joseph B. Whitney, *Megaproject: A Case Study of China's Three Gorges Project*, Armonk, NY: M. E. Sharpe, 1993.

Lynn, Frances M. and Jack D. Kartez, 'The Redemption of Citizen Advisory Committees: A Perspective from Critical Theory', in Renn, Webler and Wiedemann, eds., *Fairness and Competence in Citizen Participation: Evaluating Models for Environmental Discourse.*

MacArthur, John, 'Stakeholder Analysis in Project Planning: Origins, Applications and Refinements of the Method', *Project Appraisal*, vol. 12, no. 4, December 1997, 251–65.

McCallum, David R., 'Follow-Up to Environmental Impact Assessment: Learning from the Canadian Government Experience', *Environmental Monitoring and Assessment*, vol. 8, no. 3, 1987, 199–215.

McCarthy, S. C. and R. L. K. Tiong, 'Financial and Contractual Aspects of Build-Operate-Transfer Projects', *International Journal of Project Management*, vol. 9, no. 4, November 1991, 222–7.

McDowell, Linda, ed., *Undoing Place? A Geographical Reader*, London: Arnold, 1997.

MacEdo, Stephen, ed., *Deliberative Politics: Essays on Democracy and Disagreement*, Oxford University Press, 1999.

Mackinder, I. H. and S. E. Evans, *The Predictive Accuracy of British Transport Studies in Urban Areas*, Supplementary Report 699, Crowthorne: Transport and Road Research Laboratory, 1981.

Madsen, Bjarne, Chris Jensen-Butler and Thomas Bue Bjørner, *Transportinfrastruktur og regional udvikling: Udenlanske undersøgelser*, Transportrådet, Notat no. 93:06, Copenhagen: Danish Transport Council, December 1993.

Madsen, Bjarne, and Chris Jensen-Butler, *De regionaløkonomiske konsekvenser af en fast Femer Bælt forbindelse*, Copenhagen: AKF, 1999.

Major Projects Association, *Major Projects and the Environment*, Proceedings of a Royal Geographical Society/Major Projects Association Conference, Technical Paper no. 8, Oxford: Major Projects Association, June 1989.

Beyond 2000: A Source Book for Major Projects, Oxford: Major Projects Association, 1994.

Mandel, B., *Schnellverkehr und Modal Split*, Karlsruhe Papers in Economic Policy Research 1, Baden Baden: NOMOS, 1992.

March, James G. and Zur Shapira, 'Managerial Perspectives on Risk and Risk Taking', *Management Science*, vol. 33, no. 11, November 1987, 1404–18.

March, James G. and Johan P. Olsen, *Rediscovering Institutions: The Organizational Basis of Politics*, New York: Free Press, 1989.

Democratic Governance, New York: Free Press, 1995.

Margolis, Julius, ed., *The Analysis of Public Output*, New York: Columbia University Press, 1970.
Martinand, Claude, ed., *Private Financing of Public Infrastructure: The French Experience*, Paris: Ministry of Public Works, Transportation and Tourism, 1994.
Matthiesen, Christian Wichmann and Åke E. Andersson, *Øresundsregionen: Kreativitet, Integration, Vækst*, Copenhagen: Munksgaard, 1993.
Maxwell, James, Jennifer Lee, Forrest Briscoe, Ann Stewart and Tatsujiro Suzuki, 'Locked on Course: Hydro-Quebec's Commitment to Mega-projects', *Environmental Impact Assessment Review*, vol. 17, no. 1, 1997, 19–38.
Merewitz, Leonard, *How Do Urban Rapid Transit Projects Compare in Cost Estimate Experience?*, Reprint no. 104, Berkeley: Institute of Urban and Regional Development, University of California, 1973.
  'Cost Overruns in Public Works', in William A. Niskanen, Arnold C. Harberger, Robert H. Haveman, Ralph Turvey and Richard Zeckhauser, eds., *Benefit–Cost and Policy Analysis*, Chicago: Aldine Publishers, 1973, reprint no. 114, Berkeley: Institute of Urban and Regional Development, University of California, 1973.
Merrow, Edward W., *Understanding the Outcomes of Megaprojects: A Quantitative Analysis of Very Large Civilian Projects*, Santa Monica, CA: RAND Corporation, 1988.
Michel, G. S., D. Palsat and C. Cawley, 'Issues Relating to Performance Specifications for Roadways', paper presented at the 1998 Annual Conference of the Transportation Association of Canada.
Mills, Gordon, 'Commercial Funding of Transport Infrastructure: Lessons from Some Australian Cases', *Journal of Transport Economics and Policy*, vol. 25, no. 3, September 1991, 279–98.
Ministry of Transport, Ministry of Finance and Sund & Bælt Holding, Ltd., *Udredning af økonomien i A/S Øresundsforbindelsen (de danske anlæg)* (Copenhagen: Author, 2002).
Moe, Ronald C., 'Exploring the Limits of Privatization', *Public Administration Review*, vol. 47, no. 6, 1987, 453–60.
Mohring, Herbert, 'Maximising, Measuring and *Not* Double Counting Transportation-Improvement Benefits: A Primer on Closed- and Open-Economy Cost–Benefit Analysis', *Transportation Research*, part B, vol. 27, no. 6, 1993, 413–24.
Moles, Peter and Geoffrey Williams, 'Privately Funded Infrastructure in the UK: Participants' Risk in the Skye Bridge Project', *Transport Policy*, vol. 2, no. 2, 1995, 129–34.
Moran, Michael and Tony Prosser, eds., *Privatization and Regulatory Change in Europe*, Milton Keynes: Open University Press, 1994.
Morgan, M. Granger, Max Henrion and Mitchell Small, *Uncertainty: A Guide to Dealing with Uncertainty in Quantitative Risk and Policy Analysis*, Cambridge University Press, 1990.
Morlok, J., 'Baden Airpark and Its Importance for the High Technology Region of Karlsruhe', paper presented at Seminar on Soft Factors in Spatial Dynamics, University of Karlsruhe, February 1998.

Morone, J. G. and E. J. Woodhouse, *Averting Catastrophe: Strategies for Regulating Risky Technologies*, Berkeley: University of California Press, 1986.

Morris, Peter W. G. and George H. Hough, *The Anatomy of Major Projects: A Study of the Reality of Project Management*, New York: John Wiley & Sons, 1987.

Nash, C. A., 'Developments in Transport Policy: Rail Privatisation in Britain', *Journal of Transport Economics and Policy*, vol. 27, no. 3, 1993, 317–22.

National Audit Office, *Department of Transport, Scottish Department and Welsh Office: Road Planning*, London: HMSO, 1988.

The Private Finance Initiative: The First Four Design, Build, Finance and Operate Roads Contracts, Report by the Comptroller and Auditor-General, London: HMSO, 28 January 1998.

National Research Council, *Issues in Risk Assessment: A Paradigm for Ecological Risk Assessment*, Washington, DC: National Academy Press, 1993.

Nevitt, Peter K., *Project Financing*, London: Euromoney Publications, 1983.

Ng, Yuen Ching and W. R. Sheate, 'Environmental Impact Assessment of Airport Development Proposals in the United Kingdom and Hong Kong: Who Should Participate?', *Project Appraisal*, vol. 12, no. 1, March 1997, 11–24.

Nielsen, Kim Lynge, 'Environmental Appraisal of Large Scale Transport Infrastructure Projects', unpublished Ph.D dissertation, Aalborg: Aalborg University, Department of Development and Planning, 1998.

Nielsen, Otto Anker, Bent Flyvbjerg and Jens Rørbech, 'Notat om arbejdet med vurdering af trafikanalyser og –prognoser, Femer Bælt', paper presented at the Conference on Fehmarn Belt organised by the Ministry of Transport and the Danish Transport Council, Copenhagen, 13 January 2000.

Nijkamp, Peter and Barry Ubbels, 'How Reliable Are Estimates of Infrastructure Costs?', unpublished paper, Department of Spatial Economics, Free University, Amsterdam, undated.

Niskanen, William A., Arnold C. Harberger, Robert H. Haveman, Ralph Turvey and Richard Zeckhauser, eds., *Benefit–Cost and Policy Analysis*, Chicago: Aldine Publishers, 1973, reprint no. 114, Berkeley: Institute of Urban and Regional Development, University of California, 1973.

Novem, Wijs op Weg and Milieudefensie, *The Netherlands Travelling Clean*, Utrecht, 1994.

O'Brien, Richard, *Global Financial Integration: The End of Geography*, London: Chatham House/Pinter, 1992.

OECD, *Infrastructure Policies for the 1990s*, Paris: OECD, 1993.

Økonomigruppen (The Economy Group), 'Notat om Øresundsforbindelsens driftsøkonomiske rentabilitet', Copenhagen: Ministry of Transport, Ministry of Finance, Road Directorate and Danish State Railways, 2 October 1990.

Øresundkonsortiet, *Den faste forbindelse over Øresund*, Copenhagen: Øresundskonsortiet, 1994.

O'Riordan, Timothy, *Major Projects and the Environmental Movement*, Major Projects Association Technical Paper, no. 5, Oxford: Major Projects Association, April 1988.

'On Greening of Major Projects', in Major Projects Association, *Major Projects and the Environment*, Proceedings of a Royal Geographical Society/Major Projects Association Conference, Technical Paper no. 8, Oxford: Major Projects Association, June 1989.

Ostrom, Elinor, Larry Schroeder and Susan Wynne, *Institutional Incentives and Sustainable Development: Infrastructure Policies in Perspective*, Boulder, CO: Westview Press, 1993.

Palerm, Juan R., 'An Empirical-Theoretical Analysis Framework for Public Participation in Environmental Impact Assessment', *Journal of Environmental Planning and Management*, vol. 43, no. 5, September 2000, 581–600.

Partidário, Maria Rosário, 'Strategic Environmental Assessment: Key Issues Emerging from Recent Practice', *Environmental Impact Assessment Review*, vol. 16, no. 1, January 1996, 31–55.

Pentecost, Edwin D., 'Managing Large-Scale Environmental Impact Assessments', *Impact Assessment Bulletin*, vol. 9, no. 4, 1991, 35–40.

Phantumvanit, Dhira and Widhanya Nandhabiwat, 'The Nam Choan Controversy: An EIA in Practise', *Environmental Impact Assessment Review*, vol. 9, no. 2, June 1989, 135–47.

Pickrell, Don, *Urban Rail Transit Projects: Forecast Versus Actual Ridership and Cost*, Washington, DC: US Department of Transportation, 1990.

'A Desire Named Streetcar: Fantasy and Fact in Rail Transit Planning', *Journal of the American Planning Association*, vol. 58, no. 2, 1992, 158–76.

Pollitt, C. and H. Summa, 'Performance Auditing: Travellers' Tales', in E. Chelimsky and W. Shadish, eds., *Evaluation for the Twenty-First Century: A Resource Book*, Thousand Oaks, CA: Sage, 1997.

Powell, Walter W. and Paul J. DiMaggio, eds., *The New Institutionalism in Organizational Analysis*, Chicago: University of Chicago Press, 1991.

Preston, John, 'The Economics of British Rail Privatization: An Assessment', *Transport Reviews*, vol. 16, no. 1, 1996, 1–21.

Ramamurti, Ravi, 'Testing the Limits of Privatization: Argentine Railroads', *World Development*, vol. 25, no. 12, December 1997, 1973–93.

Randolph, John and Michael Bauer, 'Improving Environmental Decision-Making Through Collaborative Methods', *Policy Studies Review*, vol. 16, nos. 3–4, fall/winter 1999, 168–91.

Redhead, Keith and Steward Hughes, *Financial Risk Management*, Aldershot: Gower, 1988.

Regeringens proposition 1999/2000:66, *Villkoren för järnvägstrafiken på den fasta förbindelsen över Öresund m.m.*, Stockholm, n.p. 16 March 2000.

Renn, Ortwin, 'Three Decades of Risk Research: Accomplishments and New Challenges', *Journal of Risk Research*, vol. 1, no. 1, 1998, 49–71.

'A Model for an Analytic-Deliberative Process in Risk Management', *Environmental Science and Technology*, vol. 33, no. 18, September 1999, 3049–55.

Renn, Ortwin, T. Webler, H. Rakel, B. Johnson and P. Dienel, 'A Three-Step Procedure for Public Participation in Decision Making', *Policy Sciences*, vol. 26, 1993, 189–214.

Renn, Ortwin, Thomas Webler and Peter Wiedemann, eds., *Fairness and Competence in Citizen Participation: Evaluating Models for Environmental Discourse*, Dordrecht: Kluwer, 1995.

Richardson, Tim, 'The Trans-European Transport Network: Environmental Policy Integration in the European Union', *European Urban and Regional Studies*, vol. 4, no. 4, 1997, 333–46.

Richmond, Jonathan E. D., *New Rail Transit Investments: A Review*, Cambridge, MA: Harvard University, John F. Kennedy School of Government, 1998.

—— *The Private Provision of Public Transport*, Cambridge, MA: Harvard University Taubman Center for State and Local Government, 2001.

—— 'A Whole-System Approach to Evaluating Urban Transit Investments', *Transport Reviews*, vol. 21, no. 2, 2001, 141–79.

Ridley, Tony, 'The Chunnel', *Land and Infrastructure*, n. p., undated.

Rosen, B., *Holding Government Bureaucracies Accountable*, 2nd edn, New York: Praeger, 1989.

Ross, John F. L., 'High-Speed Rail: Catalyst for European Integration?', *Journal of Common Market Studies*, vol. 32, no. 2, June 1994, 191–214.

—— 'When Co-operation Divides: Øresund, the Channel Tunnel and the New Politics of European Transport', *Journal of European Public Policy*, vol. 2, no. 1, March 1995, 115–46.

—— *Linking Europe: Transport Policies and Politics in the European Union*, Westport, CT: Praeger Publishers, 1998.

Ryan, Katherine E. and Lizanne Destefano, eds., *Evaluation As a Democratic Process: Promoting Inclusion, Dialogue, and Deliberation*, San Francisco: Jossey-Bass, 2000.

Ryntveit, Geir Olav and Jon Inge Lian, *Ringvirkninger av endret veitilgjenglighet*, TØI-rapport 168, Oslo, 1993.

Sadler, Barry, *International Study of the Effectiveness of Environmental Assessment*, Interim Report and Discussion Paper, Canada: Canadian Environmental Assessment Agency, 1995.

Salazar, Joanna Gail, 'Damming the Child of the Ocean: The Three Gorges Project', *Journal of Environment and Development*, vol. 9, no. 2, June 2000, 160–74.

Sanghvi, Arun and Robert Vernstrom, 'Review and Evaluation of Historic Electricity Forecasting Experience (1960–1985)', *Industry and Energy Department Working Paper*, Energy Series Paper, no. 18, Washington, DC: World Bank, 1989.

Sankoh, O. A., 'An Evaluation of the Analysis of Ecological Risks Method in Environmental Impact Assessment', *Environmental Impact Assessment Review*, vol. 16, no. 3, May 1996, 183–8.

Savvides, Savvakis, 'Risk Analysis in Investment Appraisal', *Project Appraisal*, vol. 9, no. 1, March 1994, 3–18.

Schnaars, Steven, *Megamistakes: Forecasting and the Myth of Technological Change*, New York: Free Press, 1989.

Schulman, Paul R., *Large-Scale Policy Making*, New York: Elsevier, 1980.

Schultze, Charles L., 'The Federal Budget and the Nation's Economic Health', in Henry Aaron, ed., *Setting National Priorities: Policy for the Nineties*, Washington, DC: Brookings Institution, 1990, 19–63.

Schwass, Richard D. and Brenda Fowler, 'Public Involvement Throughout the Big Chute Hydroelectric Redevelopment Project', *Impact Assessment*, vol. 11, no. 4, 1993, 417–34.

Sclar, Elliott D., *You Don't Always Get What You Pay For: The Economics of Privatization*, Ithaca, NY: Cornell University Press, 2000.

Scott, W. Richard, *Institutions and Organizations: Theory and Research*, Thousand Oaks, CA: Sage, 1995.

Seamans, Robert C., Jr, and Frederick I. Ordway, 'The Apollo Tradition: An Object Lesson for the Management of Large-Scale Technological Endeavors', *Interdisciplinary Science Reviews*, vol. 2, no. 4, 1977, 270–304.

Seidenstat, Paul, 'Privatization: Trends, Interplay of Forces, and Lessons Learned', *Policy Studies Journal*, vol. 24, no. 3, 1996, 464–77.

Sharp, W. F., *Investments*, Englewood Cliffs, NJ: Prentice Hall, 1978.

Shepherd, Anne and Christi Bowler, 'Beyond the Requirements: Improving Public Participation in EIA', *Journal of Environmental Planning and Management*, vol. 40, no. 6, 1997, 725–38.

Sinclair, John and Alan Diduck, 'Public Education: An Undervalued Component of the Environmental Assessment Public Involvement Process', *Environmental Impact Assessment Review*, vol. 15, no. 3, 1995, 219–40.

Singleton, W. T. and Jan Hovden, eds., *Risk and Decisions*, New York: John Wiley & Sons, 1987.

Skamris, Mette K., *Large Transport Projects: Forecast Versus Actual Traffic and Costs*, Report no. 151, Aalborg: Department of Development and Planning, Aalborg University, 1994.

'Economic Appraisal of Large-Scale Transport Infrastructure Investments', unpublished Ph.D dissertation, Aalborg: Department of Development and Planning, Aalborg University, 2000.

Skamris, Mette K. and Bent Flyvbjerg, 'Inaccuracy of Traffic Forecasts and Cost Estimates on Large Transport Projects', *Transport Policy*, vol. 4, no. 3, 1997, 141–6.

Sørensen, Claus Hedegaard, *Slår bro fra kyst til kyst: en analyse af Socialdemokratiet og Øresundsforbindelsen*, Århus: Hovedland, 1993.

SOU (Statens Offentliga utredninger), *Citytunneln i Malmö*, Betänkande av utredningen med uppdrag att lämna förslag till genomförande och finansiering av en eventuell Citytunnel i Malmö, SOU 1994:78, Stockholm: SOU, 1994.

Steelman, Toddi A. and William Ascher, 'Public Involvement Methods in Natural Resource Policy Making: Advantages, Disadvantages and Trade-Offs', *Policy Sciences*, vol. 31, no. 2, 1997, 71–90.

Steer, Davies and Gleave Ltd, *Turning Trucks into Trains. The Environmental Benefits of the Channel Tunnel*, London: Transport 2000 Ltd, 1987.

Steinberg, Gerald M., 'Large-Scale National Projects as Political Symbols: the Case of Israel', *Comparative Politics*, vol. 19, no. 3, April 1987, 331–46.

Stekler, H. O., 'Are Economic Forecasts Valuable?', *Journal of Forecasting*, vol. 13, no. 6, 1994, 495–505.

Summers, Robert, 'Cost Estimates as Predictors of Actual Costs: A Statistical Study of Military Developments', in Thomas Marschak, Thomas K. Glennan and Robert Summers, eds., *Strategy for R&D: Studies in the Microeconomics of Development*, Berlin: Springer-Verlag, 1967, 140–89.

198 Bibliography

Sund & Bælt *Årsberetning 1999*, Copenhagen: Sund & Bælt Holding, 1999.
Storebælt og miljøet, Copenhagen: Sund & Bælt Holding, 1999.
'Omkostninger og fordele ved miljøprioritering', unpublished paper, 14 April 2000.
Årsberetning 2001 (Copenhagen: Sund & Bælt Holding, 2002).
Swedish Auditor-General, *Infrastrukturinvesteringar: En kostnadsjämförelse mellan plan och utfall i 15 större projekt inom Vägverket och Banverket*, RRV 1994:23, Stockholm: Avdelningen för Effektivitetsrevision, 1994.
Szyliowicz, Joseph S., 'Decision Making for Sustainable Development: Towards a New Paradigm', paper prepared for IPSA World Congress, Seoul, Korea, August 1997.
Szyliowicz, Joseph and Andrew R. Goetz, 'Getting Realistic About Megaproject Planning: The Case of the New Denver International Airport', *Policy Sciences*, vol. 28, no. 4, 1995, 347–67.
Tam, C. M., 'Build-Operate-Transfer Model for Infrastructure Developments in Asia: Reasons for Successes and Failures', *International Journal of Project Management*, vol. 17, no. 6, December 1999, 377–82.
Teichroeb, R. 'Canadian Blessing for Chinese Dam Called "Prostitution"', *Winnipeg Free Press*, 20 September 1990, p. 9.
Teigland, Jon, 'Predictions and Realities: Impacts on Tourism and Recreation from Hydropower and Major Road Developments', *Impact Assessment and Project Appraisal*, vol. 17, no. 1, March 1999, 67–76.
Thompson, Dixon and Melvin J. Wilson, 'Environmental Auditing: Theory and Applications', *Environmental Management*, vol. 18, no. 4, 1994, 605–15.
Thompson, Stewart, Joanna R. Treweek and D. J. Thurling, 'The Ecological Component of Environmental Impact Assessment: A Critical Review of British Environmental Statements', *Journal of Environmental Planning and Management*, vol. 40, no. 2, 1997, 157–71.
Time, 3 August 1998.
Tordenström, Henry, *Trafik- och näringslivseffekter av broar och andra trafiklänkar: Internationella erfarenheter av genomförda projekt*, Malmö: Stadskontoret, 1991.
Treweek, Joanna R., Stewart Thompson, N. Veitch and C. Japp, 'Ecological Assessment of Proposed Road Developments: A Review of Environmental Statements', *Journal of Environmental Planning and Management*, vol. 36, 1993, 295–307.
Tuler, S., 'Development of Mutual Understanding Among Stakeholders in Environmental Policy Disputes', in *Human Ecology: Progress Through Integrative Perspectives*, proceedings of the 7th Society for Human Ecology Meeting, East Lansing, MI: Society for Human Ecology, April 1994, 280–84.
'Learning Through Participation', *Human Ecology Review*, vol. 5, no. 1, 1998, 58–60.
Tuler, S. and T. Webler, 'Voices From the Forest: What Participants Expect of a Public Participation Process', *Society and Natural Resources*, vol. 12, no. 5, 1999, 437–53.

United Nations Economic Commission for Europe, *Post-Project Analysis in Environmental Impact Assessment*, New York: United Nations Publications, 1990.

'Konvention om vurdering af virkningerne på miljøet på tværs af Landegrænserne', Established at Espoo, Finland, 25 February 1991.

Vasseur, Liette, Lise Lafrance, Colette Ansseau, Dominique Renaud, Daniel Morin and Thérèse Audet, 'Advisory Committee: A Powerful Tool for Helping Decision Makers in Environmental Issues', *Environmental Management*, vol. 21, no. 3, 1997, 359–65.

Venables, A. J. and M. Gasiourek, 'Evaluating Regional Infrastructure: A Computable Approach', paper, London School of Economics, 1996.

Vestgaard, Jens, 'International Experience with Contractor Driven BOT Concessions. Case: A Contractor Promotes a BOT Project with MOT in a Developing Country', in Harry Lahrmann and Leif Hald Pedersen, eds., *Trafikdage på Aalborg Universitet '96*, Aalborg: Danish Transport Council and Aalborg University Transport Research Group, 1996, 419–29.

Vickerman, Roger W., 'The Channel Tunnel and Regional Development: A Critique of an Infrastructure-Led Growth Project', *Project Appraisal*, vol. 2, no. 1, March 1987, 31–40.

'Transport Infrastructure and Region Building in the European Community', *Journal of Common Market Studies*, vol. 32, no. 1, March 1994, 1–24.

'Long Term Impacts of the Channel Tunnel: Methodology and Evidence', paper for International Research Seminar on the Regional Development Impacts of the Øresund Bridge, Copenhagen, 28–29 November 1999, 11–12.

'Economic Impacts of Large Transport Infrastructure Projects', paper for Symposium on Effects of a Maglev-Train Schiphol-Groningen, Groningen, 17 November 2000.

Vickers, Geoffrey, *The Art of Judgment: A Study of Policy Making*, London: Sage, 1995; first published by Chapman and Hall, 1965.

Virilio, Paul, 'Un monde surexposé: fin de l'histoire, ou fin de la géographie?', in *Le Monde Diplomatique*, vol. 44, no. 521, August 1997, 17.

Wachs, Martin, 'Technique Vs. Advocacy in Forecasting: A Study of Rail Rapid Transit', *Urban Resources*, vol. 4, no. 1, 1986, 23–30.

'When Planners Lie with Numbers', *Journal of the American Planning Association*, vol. 55, no. 4, autumn 1989, 476–9.

'Ethics and Advocacy in Forecasting for Public Policy', *Business and Professional Ethics Journal*, vol. 9, nos. 1–2, 1990, 141–57.

Wallin, Bruce A., 'The Need for a Privatization Process: Lessons from Development and Implementation', *Public Administration Review*, vol. 51, no. 1, 1997, 11–20.

Ward, Hugh, 'Citizens' Juries and Valuing the Environment: A Proposal', *Environmental Politics*, vol. 8, no. 2, Summer 1999, 75–96.

Webler, Thomas, ' "Right" Discourse in Citizen Participation: An Evaluative Yardstick', in Renn, Webler and Wiedemann, eds., *Fairness and Competence in Citizen Participation: Evaluating Models for Environmental Discourse*.

'Organizing Public Participation: A Review of Three Handbooks', *Human Ecology Review*, vol. 3, no. 1, 1997, 245–54.

Webler, Thomas, H. Rakel and R. J. S. Ross, 'A Critical Theoretic Look at Technical Risk Analysis', *Industrial Crisis Quarterly*, vol. 6, 1992, 23–38.

Webler, Thomas and Seth Tuler, 'Fairness and Competence in Citizen Participation: Theoretical Reflections From a Case Study', *Administration and Society*, vol. 32, no. 5, November 2000, 566–95.

Weeks, Edward C., 'The Practice of Deliberative Democracy: Results from Four Large-Scale Trials', *Public Administration Review*, vol. 60, no. 4, July–August 2000, 360–72.

Wells, Sarah Stewart and Bruce G. Hutchinson, 'Impact of Commuter-Rail Services in Toronto Region', *Journal of Transportation Engineering*, vol. 122, no. 4, July/August 1996, 270–75.

White, Fidelma and Kathryn Hollingsworth, *Audit, Accountability and Government*, Oxford: Clarendon Press, 1999.

Whitworth, Alan and Christopher Cheatham, 'Appraisal Manipulation: Appraisal of the Yonki Dam Hydroelectric Project', *Project Appraisal*, vol. 3, no. 1, March 1988, 13–20.

Williams, Walter, *Honest Numbers and Democracy*, Washington, DC: Georgetown University Press, 1998.

Winklhofer, Heidi, Adamantios Diamantopoulos and Stephen F. Witt, 'Forecasting Practice: A Review of the Empirical Literature and an Agenda for Future Research', *International Journal of Forecasting*, vol. 12, no. 2, 1996, 193–221.

Winnipeg Free Press, 20 September 1990.

Winston, Clifford, 'Efficient Transportation Infrastructure Policy', *Journal of Economic Perspective*, vol. 5, no. 1, winter 1991, 113–27.

Wood, Christopher, 'Lessons from Comparative Practice', *Built Environment*, vol. 20, no. 4, 1994, 332–44.

Wood, Christopher, Ben Dipper and Carys Jones, 'Auditing the Assessment of the Environmental Impacts of Planning Projects', *Journal of Environmental Planning and Management*, vol. 43, no. 1, January 2000, 23–47.

Wood, Graham, 'Post-development Auditing of EIA Predictive Techniques: A Spatial Analytical Approach', *Journal of Environmental Planning and Management*, vol. 42, no. 5, September 1999, 671–89.

World Bank, *An Overview of Monitoring and Evaluation in the World Bank*, Report no. 13247, Washington, DC: Operations Evaluation Department, World Bank, 1994.

Evaluation Results 1992, Washington, DC: Operations Evaluation Department, World Bank, 1994.

World Development Report 1994: Infrastructure for Development, Oxford: Oxford University Press, 1994.

Economic Analysis of Projects: Towards a Results-Oriented Approach to Evaluation, ECON Report, Washington, DC: World Bank, undated.

Wright, Vincent, ed., *Privatization in Western Europe: Pressures, Problems and Paradoxes*, London: Pinter Publishers, 1994.

Zahariadis, Nikolaos and Christopher S. Allen, 'Ideas, Networks, and Policy Streams: Privatization in Britain and Germany', *Policy Studies Review*, vol. 14, nos. 1–2, spring–summer 1995, 71–98.

Index

Aalborg transport infrastructure study
    cost overruns 15–16, 18, 23
    forecast and actual traffic 26–28
    urban rail projects 37–38
access links 118–120
accountability 107–124
    basic instruments 110–111
    conflicts of interest 138–139
    decision making 7, 125–135, 138, 141
    Fehmarn Belt link case study 143–151
    government 140–141
    lack of 137–138
    performance specifications 115–117, 123–124, 139–140, 147–148
    regulatory regimes 118–120, 124, 140, 148
    risk capital 120–124, 140–141
    transparency 108, 111–115, 123, 139, 146–147
    viability overestimation 44
airports 3–4, 27
allocation of risk 83, 94–95, 97
Americas, megaproject examples 1–2
Apollo aerospace programme 76
appraisal
    consultant bias 30
    optimism 73, 137
    project promoter bias 31
    see also conventional project development; project development and appraisal
Asian megaproject examples 1, 2
Auditor-General of Denmark 75
Auditor-General of Sweden transport infrastructure study 15, 42
audits
    environmental 55–57, 63–64, 149–150
    see also post-auditing

Baltic Sea
    environmental impact assessment 51–52
    extended peer review 113
    Fehmarn Belt link 113, 143–151
    Great Belt link 58–60
    Zero Solutions 58–60, 122–123, 149
Basic Policy Documents 130
Beck, Ulrich 6
benefit misrepresentation 46–48
bias
    consultants 30
    project promoters 31
boundaries, national 70–71
bridges
    Great Belt link 58
    private-sector involvement 93, 95
    traffic forecasts 101–102
    urban area connection 70–71
build-operate-transfer (BOT) approach
    accountable decision making 125, 127–128, 141
    concept 93–94
    financing characteristics 95
        ingredients for success 104
        long-term (operations) interests 97
build-own-operate-transfer (BOOT) projects 93, 96
Bundesverkehrswegeplan 50
bustards 54–55

canals 18
capacity problems 119–120
capital
    risk 120–124, 140–141
    state-owned enterprise mobilisation 129–130
capital-market risks 77
carbon dioxide emissions 60–61
Channel tunnel
    concession financing 96–97
    cost overrun 3, 12, 14, 19–20, 96
    development effects 68–69
    environmental impact assessment 50–51
    Eurotunnel 32–34, 74, 96–97
    risk analysis 74

201

Channel tunnel (*cont.*)
  traffic forecasts 22–23, 25
  viability 32–34
checks and balances 20, 44, 108
China 99
Chunnel *see* Channel tunnel
citizen participation 111–112
  see also public participation
communication, transparency 111–112
company approach 133
competition
  public-sector accountability 108–109
  tendering 117
  transport infrastructure 66–67
complementary factors, prediction failure 29
complementary investments 118, 140
concerned groups
  planning involvement 89
  see also public participation; stakeholders
concessions
  accountable decision making 125–128
  Channel tunnel 96–97
  infrastructure funding 93–95
  investment 78
  toll roads 98–99
  Warnow tunnel 102–103
conflicts of interest 90–91, 138–139
congestion, Øresund link 119–120
consultants 30
Consultation Documents 1 and 2 131–132
contractors
  lack of accountability 45
  risk allocation 83
  see also special interest groups
contracts
  design-build-finance-operate (DBFO) 97, 102–103
  Eurotunnel 96–97
  infrastructure provision 81, 103
  input-based 103
  performance specification 103, 117
  project financing 94–95
conventional project development 86–91
  alternative approach comparison 133–134
  governmental roles 90–91
  institutional issues 87–88, 91
  introduction 86–88
  public participation 87–90
  role of government 138–139
  stakeholder involvement 87–89
  steps 87
  technical solutions 86, 89–90

cost
  bustard lives 54–55
  Channel tunnel 96
  environmental protection 62
  market risks 84
  misrepresentation 46–48
  project-specific risks 83
  risk 77–80
  transport infrastructure 66, 71, 118–120
cost overruns
  Channel tunnel 96
  Great Belt link 75
  history 11–21
  inflated viability 44
  megaproject performance 3–4
  Øresund link 75
  private sector 95
  risk 76–77
  urban rail projects 38
cost–benefit analyses
  economic growth effects 67–68
  German environmental impact assessment 53
  megaproject preparation 5
  tactical estimation 137
  untrustworthiness 20

Danish Transport Council 143–144, 150
data quality, prediction failure 29
deception 46–48
Decision Documents 132
decision making
  accountability 7, 125–135, 138, 141
  democracy 5, 10
  environmental impact assessment 50
  French transparency 114
  institutional arrangements 138
  performance specifications 139–140
  risk 138
  risk assessment 6–7
  special-interest group capture 88, 134
  stakeholders 113
delusion 46–48
demand forecasts 22–31
  Great Belt link 61
  inflated viability 44–45
  Øresund link 71
  toll roads 100–102
democracy 142
democracy deficit 5, 115
Denmark
  environmental impact assessment 50, 52
  Fehmarn Belt link case study 143–151
  fixed link appraisal 122–123
  joint-stock companies 129–130

Index

railways 119–120
  see also Great Belt link; Øresund link
design-build-finance-operate (DBFO) projects 97, 102–103
Deutsche Bahn AG 39–41
discontinuous behaviour, prediction failure 29
documentation, project development process 130–132
downside probability neglect 80
Draft Performance Specifications Reports 130

economic growth 65–72
  Fehmarn Belt link 150
  investments 78
  road transport demand 100
  transport infrastructure 65–72
economic risks 133–134
economic rules 140
EGAP-principle (Everything Goes According to Plan) 80
EIA (environmental impact assessment) 49–64, 149–150
elasticity 100, 105
energy consumption 60–61
environmental audits 55–57, 63–64, 149–150
environmental impact assessment (EIA) 49–64, 149–150
environmental learning 49, 62–64
equity, state-owned enterprise capital mobilisation 129–130
estimates
  overruns 11–21
  short-term versus long-term interests 137, 149
ethics 48
European megaproject examples 1
European Union 30
Eurotunnel
  Channel tunnel viability 32–34
  private-sector involvement 96–97
  risk analysis 74
  see also Channel tunnel
exogenous factors, prediction failure 29
expertise 112–113
extended peer review 63, 112–115

feasibility studies
  extended peer review 113–115
  predictability assumption 73–74
  risk analysis 84–85
  risk assessment 80–81
  risk management 81–84

Fehmarn Belt link case study 143–151
Final Performance Specification Document 132
financial risk 76–77
financing, private sector 92–95, 120–121, 141
fixed connections see Fehmarn Belt link case study; Great Belt link; Øresund link
force-majeure risks 77, 83
forecasts
  deceptive 46–48
  demand 22–31
  Great Belt link 61
  Øresund link 71
  toll roads 100–102
France
  toll roads 98–99, 101
  transparency 114
freight transport
  Channel tunnel forecast 22–23
  Germany 24
  modelling 28–29
frictionless society 2–3

Germany
  decision-making evaluation 50
  demand forecasting 24, 29
  environmental impact assessment 53–55
  Fehmarn Belt link case study 143–151
  high-speed rail projects 38–41, 54–55
  road projects 102–103
Giddens, Anthony 6
good practice
  risk analysis 76
  transport infrastructure projects 21
government
  accountability 134
  civil society groups 111–112
  commitment to build-operate-transfer approach 128
  conflicts of interest 90–91, 138–139
  conventional project development 90–91, 138–139
  inability to enforce accountability 120, 140–141
  public interest protection 111–112, 123, 134
  public-sector involvement 109–110
  public–private partnerships 102, 106
  stakeholders 111–112
Great Belt link
  capacity problems 119
  common mussels 59
  cost overrun 12–14, 19

Great Belt link (*cont.*)
  development effects 69
  environmental impact 149–150
  environmental issue downplay 4
  institutional arrangement 108–109
  polarisation of debate 89
  post-auditing 57–62
  risk analysis 74–75, 149
  sovereign guarantees 79
  traffic forecasts 23–24
  viability 34–36
Great Belt, Ltd 34–36
growth
  economic 65–72, 78, 100, 150
  regional 65–72, 150
guarantees
  sovereign 79, 85, 109, 120, 145–146
  state-owned enterprise capital mobilisation 129–130

high-speed rail projects 38–41, 54–55
honesty 43–44
Hong Kong 3–4
Hungary 100

imperfect markets 67–68
inertia, Fehmarn Belt link 147
inflated viability 44–46
information
  relevant 84
  transparency 111–115
Information Documents 132
infrastructure
  concept 2–3
  contracting 81
  investment 65–72, 78
  organisational structure 121–122
  private-sector involvement 92–106
  regional development and economic growth 65–72
  risk 78, 84
  see also transport infrastructure
institutional issues
  conventional project development 87–88, 91
  good decision making 138
  Great Belt link 108–109
  Øresund link 108–109
  public and private responsibilities 109–110
integration across national boundaries 70–71
international development finance 90
Internet 2
investment

complementary 118, 140
risk capital 120–123
risks 78–79
transport infrastructure 65–72

joint-stock companies 108, 129–130

land availability 66–67
leadership, public *versus* private 107
learning, environmental 49, 62–64
legislation 134–135
lies 47
lobby groups *see* special interest groups
long-term (operations) interests 97, 128

M1–M15 toll motorway 100
M62 motorway study 71
MAGLEV rail project 38–39
Malmö 69–70
market forces 121
market imperfections 67–68
market risks 77, 84, 100–101, 105
metro (rail transit) systems
  Transport and Road Research Laboratory study 15, 26, 43
  US Department of Transport study 15, 25, 42–43
misinformation 48
  see also deception
MLD principle (Most Likely Development) 80, 84
modelling, forecasting uncertainty 28–29
monitoring 63–64
motorways 71, 98–100
mussels 59
Myanmar–Thailand gas pipeline 9–10

national boundary integration 70–71
negligence, risk 137–138

optimism
  appraisal 73, 137
  Great Belt link 74–75
Øresund link
  access link costs 118–119
  congestion 119–120
  cost overrun 12–14, 19, 75
  development effects 69–71
  downplay of environmental issues 4
  environmental impact 51–52, 149–150
  institutional arrangement 108–109
  lack of risk analysis 149
  performance specification 117
  polarisation of debate 89

Index

post-auditing 57–62
risk analysis 75
risk costs 80
sovereign guarantees 79
traffic 24, 61–62, 71
viability 36–37
ownership 96–97

paradox, megaprojects
　causes 137–138
　cures 138–141
　introduction 1–10
　overview 136
Pareto improvements 122–123
participation, public *see* public participation
peer reviews
　environmental impact assessment 63
　transparency 112–115
performance
　contracting 103
　criteria 133
　demand forecasting 31
　management 141
　megaprojects 3–5, 136
performance specifications
　accountability 115–117, 123–124, 139–140, 147–148
　Basic Policy Documents 130
　Consultation Document 1 131
　contracts 103, 117
　decision making 132, 139–140
　Draft Reports 130
　Fehmarn Belt link 147–149
　Final Document 132
　Øresund link 117
　policy objectives 115–116
　safety 115–116
　stakeholders 116
pipelines 9–10
planning
　basic parameter establishment 89–90
　concerned group involvement 89
polarisation of debate 89
policy objectives 115–116
politics
　lack of accountability 45
　polarisation 89
　prediction failure cause 30
post-auditing 62–64, 150
power relations 7
Pre-Feasibility Study Documents 130–131
predictions
　environmental impacts 55–57
　failures 28–31

price
　regulation 118
　transport infrastructure investment 66, 71
private sector 92 106
　borderline with public sector 9, 107–110, 138–139
　design-build-finance-operate projects 97, 102–103
　Eurotunnel 96–97
　Fehmarn Belt link 144–146, 148
　institutional issues 109–110
　lessons 104–106
　megaproject financing 92–95, 120–121, 141
　participation without sovereign guarantee 120–121, 141
　risk spreading 83
　toll roads 98–102
project development and appraisal
　concession approach 125–128
　conventional and alternative comparison 133–134
　Fehmarn Belt link case study 143–151
　legislation 134–135
　process documentation 130–132
　state-owned enterprise approach 125, 127–130
　see also conventional project development
project promoters 31, 46–48, 137–138, 149
project-specific risks 77, 83
protest actions 83
public attitudes 5
public participation
　conventional project development 87–90
　Fehmarn Belt link 146–147
　transparency 111–112, 139
public sector
　borderline with private sector 9, 107–110, 138–139
　conventional approach 86
　Fehmarn Belt link 144–145
　infrastructure development 92
　institutional issues 109–110
public works acts 135
public–private partnerships 102, 106

rail projects
　associated costs 118–120
　Channel tunnel forecasts 22–23
　demand forecasting 24–26, 31
　environmental impact assessment 51
　German high-speed 38–41, 54–55

rail projects (*cont.*)
  Great Belt link 23
  Sweden 15
  urban 37–38
rail transit systems *see* metro (rail transit) systems
regional growth effects 65–72, 150
regulatory regimes 140
  accountability 118–120, 124
  Fehmarn Belt link 148
relevant information 84
rent-seeking behaviour 121, 134, 137–138, 149
risk
  allocation 83, 94–95, 97
  cost 77–80, 84
  decision making 138
  deliberation 6–7
  elimination 77–78, 82–83
  environmental impact 49–64
  Fehmarn Belt link case study 143–151
  lessons 84–85
  linked to megaproject idea 9
  overrun 11–12
  regional and economic growth effects 72
  spreading 83
  toll roads 100–101
  typology 76–77
risk analysis 73–76, 84–85
  Fehmarn Belt link 148–149
  good decision making 138
risk assessment 80–81
risk capital 120–124, 140–141
risk management 81–85
  Consultation Document 2 131–132
  environmental issues 57
  German high-speed rail 39–41
  good decision making 138
risk negligence 137–138
risk society 6, 10, 142
road projects
  demand forecasting 24–28, 31
  environmental impact assessment 52–53
  Germany 102–103
  Great Belt link 23–24, 60–61
  Sweden 15
  toll roads 95, 98–102
roles
  definition 88, 90–91
  public and private sectors 107–110
rules, economic 140

safety performance specifications 115–116
scenario forecasts 30
scrutiny, public *see* transparency

SEA *see* strategic environmental assessment
sector-policy risks 77
sensitivity analysis 73
services, transport infrastructure as input 92–93
shadow tolls 103
short-term (construction) interests 97, 137–138, 149
Skye Bridge project 95
socio-economic effects
  Fehmarn Belt link 150
  lack of development phase consideration 4–5
SOE *see* state-owned enterprises
sovereign guarantees 79, 85, 109, 120, 145–146
space, independence from 2–3, 136
Spain, toll roads 98–99
spatial interaction 70–71
special interest groups
  decision-making process capture 88, 134
  lack of accountability 45
  rent-seeking behaviour 121, 134
spreading risk 8
stakeholders
  conventional project development 87–89
  decision making 113
  environmental impact assessment 63
  performance specifications 116
  public participation 139
  risk assessment 6–7
  transparency 111–112
  underestimation of costs 46
state-owned enterprises (SOE)
  accountable decision making 141
  project development 125, 127–130
  risk capital 121–122
stocks and shares 32–33
strategic environmental assessment (SEA) 55
Suez canal 18–19
Sund & Bælt, Great Belt link 58–61
Sweden
  Auditor-General transport infrastructure study 15, 42
  see also Great Belt link; Øresund link
systems analysis 55

technical solutions 86, 89–90
tendering, competitive 117
Thailand 9–10
time availability 66–67
time overruns 95
toll roads 95, 98–102, 105

Index

traffic forecasts 22–31, 100–102
Trans-European Transport Network (TTEN) 8, 55
transit systems *see* metro system
transparency
  accountability 108, 111–115, 123, 139
  Fehmarn Belt link 146–147
  France 114
  state-owned enterprises 128–129
Transport Council *see* Danish Transport Council
transport infrastructure
  associated costs 118–120
  economic development 65–72
  forecast *vs* actual cost 15–16
  good practice 21
  private-sector involvement 20, 92–106
see also infrastructure
Transport and Road Research Laboratory metro system study 15, 26, 43
TTEN (Trans-European Transport Network) 8, 55
tunnels
  Great Belt link 58, 59
  private-sector involvement 93, 102–103
  Warnow crossing 102–103
see also Channel tunnel

United Kingdom
  design-build-finance-operate projects 97, 102–103
  environmental impact assessment 52–53
  traffic forecasting 24–25
  urban area connection 70–71
  urban rail projects 37–38
US Department of Transport rail project study 15–25, 42–43

viability
  Auditor-General of Sweden transport infrastructure study 42
  Channel tunnel 32–34
  downside probability neglect 80
  feasibility study documents 130–131
  Fehmarn Belt link 143–151
  fixed-link appraisal 122–123
  forecast/actual difference 32, 43, 45–46
  Great Belt link 34–36
  inflated 44–46
  key variables 32
  metro systems 43
  Øresund link 36–37, 75
  Pre-Feasibility Study Documents 130–131
  projects with problems 42
  rail transit projects 42–43
  urban rail projects 37–38

Wachs, Martin 46–47
Wapenhans report 43
Warnow crossing tunnel 102–103
wildlife 54–55
World Bank
  EGAP-principle 80
  sovereign guarantees 79
  Wapenhans report 43
worst-case scenario analysis 81, 85

Zero Solution
  Baltic Sea environmental impacts 122–123
  Fehmarn Belt link 149
  Great Belt link 58–60
zero-friction society 2–3